DEVELOPMENTS IN MATHEMATICAL AND EXPERIMENTAL PHYSICS

Volume B: Statistical Physics and Beyond

DEVELOPMENTS IN MATHEMATICAL AND EXPERIMENTAL PHYSICS

Volume B: Statistical Physics and Beyond

Edited by

Alfredo Macias,
Francisco Uribe, and
Enrique Diaz

Universidad Autónoma Metropolitana–Iztapalapa
Mexico City, Mexico

Springer Science+Business Media, LLC

Proceedings of the First Mexican Meeting on Mathematical and Experimental Physics, held September 10–14, 2001, in Mexico City, Mexico

ISBN 978-1-4613-4965-5   ISBN 978-1-4615-0207-4 (eBook)
DOI 10.1007/978-1-4615-0207-4
©2003 Springer Science+Business Media New York
Originally published by Kluwer Academic / Plenum Publishers, New York in 2003
Softcover reprint of the hardcover 1st edition 2003

http://www.wkap.com

10 9 8 7 6 5 4 3 2 1

A C.I.P. record for this book is available from the Library of Congress

This book is dedicated to the 70th birthday of Leopoldo García–Colín
and to the 80th birthday of Nicholas van Kampen

Contributing Authors

- O. Alarcón-Waess
  Departamento de Física y Matemáticas, UDLA, Puebla,
  Santa Catarina Martir, Cholula 72820 Puebla, Mexico.

- José Luis Arauz-Lara
  Instituto de Física "Manuel Sandoval Vallarta", Universidad Autónoma
  de San Luis Potosí,
  Alvaro Obregón 64, 78000 San Luis Potosí, SLP, México.

- J. F. J. Alvarado
  Chemical Engineering Department, Instituto Tecnologico de Celaya,
  Celaya, Guanajuato, 38010 México.

- E. Ávalos
  Departamento de Física, Universidad Autónoma Metropolitana
  Av. San Rafael Atlixco 186, Col. Vicentina, Iztapalapa, México,
  D.F., 09340 México.

- Rubén G. Barrera
  Instituto de Física, Universidad Nacional Autónoma de México,
  Apartado Postal 20-364, 01000 México D. F., México.

- Ana Laura Benavides
  Instituto de Física, Universidad de Guanajuato,
  León, Guanajuato, México.

- J. F. Camacho
  Instituto de Física, Universidad Nacional Autónoma de México,
  Apdo. Postal 20-364, 01000 México, D.F., México.

- M. D. Carbajal-Tinoco
  Depto. de Física, CINVESTAV-IPN,
  Av. IPN 2508, Col. San Pedro Zacatenco, 07360 México D. F.,
  México.

- R. Castañeda-Priego
  Depto. de Física, CINVESTAV-IPN,
  Av. IPN 2508, Col. San Pedro Zacatenco, 07360 México D. F.,
  México.

- F. del Río
  Departamento de Física, Universidad Autónoma Metropolitana

Av. San Rafael Atlixco 186, Col. Vicentina, Iztapalapa, México, D.F., 09340 México.

- Andrés F. Estrada-Alexanders
  Departamento de Física, Universidad Autónoma Metropolitana
  Av. San Rafael Atlixco 186, Col. Vicentina, Iztapalapa, México, D.F., 09340 México.

- P. González-Mozuelos
  Depto. de Física, CINVESTAV-IPN,
  Av. IPN 2508, Col. San Pedro Zacatenco, 07360 México D. F., México.

- Augusto García-Valenzuela
  Centro de Instrumentos, Universidad Nacional Autónoma de México,
  Apartado Postal 70-18, 04510 México D. F., México.

- Alejandro Gil-Villegas
  Instituto de Física, Universidad de Guanajuato,
  León, Guanajuato, México.

- L. L. Goncalves
  Departamento de Física, Universidade Federal do Ceará, Campus do Pici,
  C. P. 6030, 60451-970, Fortaleza, Ceará, Brazil.

- Christian Gorba
  Max Planck Institute of Biophysics,
  Kennedyallee 70, 60596 Frankfurt, Germany.

- E. Haro-Poniatowski
  Universidad Autónoma Metropolitana Iztapalapa, Departamento de Física,
  Apdo postal 55-534, Col. Vicentina, 09340 México D.F., México.

- Volkhard Helms
  Max Planck Institute of Biophysics, Kennedyallee 70,
  60596 Frankfurt, Germany.

- M. Hernández-Contreras
  Departamento de Física, Centro de Investigación y Estudios Avanzados del Instituto Politécnico Nacional
  A.P. 14-740, México D.F., México.

- J. Hernández-Rosas
  Universidad Autónoma Metropolitana Iztapalapa, Departamento

de Física,
Apdo postal 55-534, Col. Vicentina, 09340 México D.F., México.

- Niza Ibarra-Avalos
  Instituto de Física, Universidad de Guanajuato,
  León, Guanajuato, México.

- J. I. Jiménez-Aquino
  Departamento de Física, Universidad Autónoma Metropolitana,
  Iztapalapa
  Apdo. Postal 55–534, México, D. F., 09340 México.

- M. Jouanne
  Université Pierre et Marie Curie, LMDH, UMR 7603
  4 Place Jussieu, 75252 Paris Cedex 05, France.

- Daimler Justo
  Departamento de Física, Universidad Autónoma Metropolitana
  Av. San Rafael Atlixco 186, Col. Vicentina, Iztapalapa, México,
  D.F., 09340, México.

- M. Kanehisa
  Université Pierre et Marie Curie, LMDH, UMR 7603
  4 Place Jussieu, 75252 Paris Cedex 05, France.

- Markus Lill
  Max Planck Institute of Biophysics, Kennedyallee 70,
  60596 Frankfurt, Germany.

- M. López de Haro
  Centro de Investigación en Energia, U.N.A.M.
  Apdo. Postal 34, Temixco, Mor. 62580, México.

- G. Luna-Bárcenas
  Laboratorio de Investigación en Materiales, CINVESTAV Unidad
  Querétaro, Querétaro, 76230 México.

- Antonio Martínez-Richa
  Facultad de Química, Universidad de Guanajuato,
  Guanajuato, Guanajuato, México.

- José L. Mateos
  Instituto de Física, Universidad Nacional Autónoma de México,
  Apartado Postal 20-364, 01000 México, D.F., México.

- M. Mayorga
  Facultad de Ciencias, Universidad Autónoma del Estado de México
  Av. Instituto Literario 100, Toluca, 50000 México.

- J. M. Méndez-Alcaraz
  Depto. de Física, CINVESTAV-IPN,
  Av. IPN 2508, Col. San Pedro Zacatenco, 07360 México D. F.,
  México.

- J. F. Morhange
  Université Pierre et Marie Curie, LMDH, UMR 7603
  4 Place Jussieu, 75252 Paris Cedex 05, France.

- C. Ortíz-Estrada
  Chemical Engineering Department, Instituto Tecnologico de Celaya,
  Celaya, Guanajuato, 38010 México.

- Michel Picquart
  Universidad Autónoma Metropolitana Iztapalapa, Departamento
  de Física,
  Apdo postal 55-534, Col. Vicentina, 09340 México D.F., México.

- G. Ramírez-Santiago
  Instituto de Física, UNAM, PO Box 20-364, México 01000 D. F.,
  Mexico,
  Chemical Engineering Department, The University of Texas at
  Austin, Austin TX 78712 USA.

- Angeles Ramírez-Saito
  Instituto de Física "Manuel Sandoval Vallarta", Universidad Autónoma
  de San Luis Potosí,
  Alvaro Obregón 64, 78000 San Luis Potosí, SLP, México.

- R. F. Rodríguez
  Instituto de Física, Universidad Nacional Autónoma de México,
  Apdo. Postal 20-364, 01000 México, D.F., México.

- A. Rodríguez-López
  Depto. de Física, CINVESTAV-IPN,
  Av. IPN 2508, Col. San Pedro Zacatenco, 07360 México D. F.,
  México.

- M. Romero-Bastida
  Departamento de Física, Universidad Autónoma Metropolitana,
  Iztapalapa
  Apdo. Postal 55–534, México, D. F. 09340 México.

- Víctor Romero-Rochín
  Instituto de Física, Universidad Nacional Autónoma de México.
  Apartado Postal 20-364, 01000 México, D.F, México.

- Subir Sachdev
  Department of Physics, Yale University,
  P.O. Box 208120, New Haven CT 06520-8120, USA.

- I. C. Sanchez
  Chemical Engineering Department,
  The University of Texas at Austin, Austin TX 78712 USA.

- J. Tagüeña-Martínez
  Centro de Investigación en Energia, U.N.A.M.
  Apdo. Postal 34, Temixco, Mor. 62580, México.

- Paul H.E. Tiesinga
  Sloan-Swartz Center for Theoretical Neurobiology and Computional
  Neurobiology Lab, Salk Institute
  10010 N. Torrey Pines Rd, La Jolla, CA 92037, USA.

- A. P. Vieira
  Instituto de Física, Departamento de Física Geral Universidade de
  São Paulo
  C.P. 66318, 05315-970, São Paulo, SP, Brazil.

Preface

The FIRST MEXICAN MEETING ON MATHEMATICAL AND EXPERIMENTAL PHYSICS was held at EL COLEGIO NACIONAL in Mexico City, Mexico, from September 10 to 14, 2001. This event consisted of the LEOPOLDO GARCÍA–COLÍN SCHERER Medal Lecture, delivered by Prof. Nicholas G. van Kampen, a series of plenary talks by Leopoldo García–Colín, Günter Nimtz, Luis F. Rodríguez, Rubén Barrera, and Donald Saari, and of three parallel symposia, namely, Cosmology and Gravitation, Statistical Physics and Beyond, and Hydrodynamics and Dynamical Systems. The response from the Physics community was enthusiastic, with over 200 participants and around 80 speakers, from all over the world: USA, Canada, Mexico, Germany, France, Holland, United Kingdom, Switzerland, Spain, and Hungary.

The main aim of the conference is to provide a scenario to Mexican researchers on the topics of Mathematical and Experimental Physics in order to keep them in contact with work going on in other parts of the world and at the same time to motivate and support the young and mid–career researchers from our country. To achieve this goal, we decided to invite as lecturers the most distinguished experts in the subjects of the conference and to give the opportunity to young scientist to communicate the results of their work. The plan is to celebrate this international endeavor every three years.

The most outstanding researcher at Universidad Autonoma Metropolitana is Leopoldo García–Colín. His devotion to science and high level of energy and enthusiasm that he brings to his research and teaching are very much appreciated by his students and collaborators. Therefore, the Universidad Autónoma Metropolitana (UAM) instituted in 2001 the **Leopoldo García–Colín Medal**, which will be awarded every three years. The medal is given in his honor regarded as one of the greatest mexican physicist of the 20th century and a Distinguished Professor of our University. The **Leopoldo García–Colín Medal** award recognizes outstanding scientists, in the developing and advanced countries, who have made outstanding contributions to theoretical physics. It is the highest recognition accorded by the UAM for excellence in scientific research to whom has significantly contributed to the advancement of science. The main goal of this award is to promote the academic activities at UAM by means of the wisdom of the medal winners.

An international committee of distinguished scientist selects the winners from a list of nominated candidates. The committee invites nominations from scientists working in the field of Theoretical Physics.

The award consists of a gold medal, which displays Leopoldo García–Colín in a side pose, as well as a plaque on which major contributions of the award winner are mentioned. The selection of the awardees is made solely on scientific merit.

The first awardee is Prof. Nicholas van Kampen from Utrecht University in Netherlands, for his outstanding work which has significantly contributed to the advancement of science, for the inspirational quality of his research, his thoughtful guidance of graduated and undergraduate students, his graciousness as a colleague and his service to the scientific community.

The proceedings of the FIRST MEXICAN MEETING ON MATHEMATICAL AND EXPERIMENTAL PHYSICS consist of three volumes, namely, Volume A: Cosmology and Gravitation, Volume B: Statistical Physics and Beyond, and Volume C: Hydrodynamics and Dynamical Systems.

These three volumes contain lecture notes on the topics covered in each of the three symposia at this conference. Additionally, the Proceedings are dedicated to honor the outstanding contributions to science of Professor Leopoldo García–Colín on the occasion of his 70th birthday, and of Professor Nicholas G. van Kampen on the occasion of his 80th birthday.

We would like to thank everyone who contributed to the success of the 1st MEXICAN MEETING ON MATHEMATICAL AND EXPERIMENTAL PHYSICS. Very special thanks are due to the invited speakers and lecturers who delivered a very interesting set of talks and shared their knowledge and time with participants. We also thank young people just starting out on their careers.

The Meeting could not have been realized without the financial support of El Colegio Nacional, of CONACyT (Mexico), of the Germany–Mexico exchange program of the DLR (Bonn)–CONACYT (Mexico City), and of Silicon Graphics. We wish to thank Dr. José Luis Gázquez, Dr. Luis Mier y Terán, Dr. María José Arroyo, General Rector, Rector of the Campus Iztapalapa, and Dean at Iztapalapa of the Faculty of Basic Sciences and Engineering of the Universidad Autónoma Metropolitana, respectively, for sponsoring this international and multidisciplinary endeavor. We also thank Juan Azorín from the Mexican Physical Society for his help and support.

We thank specially Prof. Leopoldo García–Colín and all the staff of EL COLEGIO NACIONAL for the warm hospitality that was extended to all participants.

We hope this Proceedings will be of interest to all the participants, and indeed to mathematicians and physicists in general, specially to young people just starting their scientific careers.

ALFREDO MACIAS, ENRIQUE DIAZ, FRANCISCO J. URIBE

Preface to the Volume B: Statistical Physics and Beyond

The FIRST MEXICAN MEETING ON MATHEMATICAL AND EXPERIMENTAL PHYSICS was held at EL COLEGIO NACIONAL in Mexico City, Mexico from September 10 to 14, 2001.

These Proceedings contain lecture notes on the topics covered during the SYMPOSIUM ON STATISTICAL PHYSICS AND BEYOND, the contributions cover the invited talks as well as the short talks. The short talks provided a pleasant and efficient way to learn about research work being developed by the participants and bring people together.

We would like to thank everyone who contributed to the success of the Symposium on Statistical Physics and Beyond. Very special thanks are due to the invited speakers and lecturers who gave a very interesting and high quality set of talks and shared their knowledge and time with participants.

The Symposium on Statistical Physics and Beyond could not have been realized without the finantial support of the CONACyT (Mexico), of the Germany–Mexico exchange program of the DLR (Bonn)–CONACYT (Mexico City), and of Silicon Graphics. We wish to thank Dr. José Luis Gázquez, Dr. Luis Mier y Terán, Dr. María José Arroyo, Rector, Rector of the campus Iztapalapa, Dean of the Faculty of Basic Sciences and Engineering of the Universidad Autónoma Metropolitana, respectively, for sponsoring this international and multidisciplinary endeavour. We also thank Dr. Juan Azorín form the Mexican Physical Society for his help and support.

We thank specially Prof. Leopoldo García–Colín and all the crew at EL COLEGIO NACIONAL for the warm hospitality that was extended to all participants.

We hope that these Proceedings will serve to further the impressive growth of Statistical Physics in the region and reinforce the already strong ties between Mexican scientists and scientist in other parts of the world.

ENRIQUE DIAZ-HERRERA

Contents

Contents

I

COLLOIDAL PARTICLES AND STOCHASTIC DYNAMICS

INTERACTION BETWEEN COLLOIDAL PARTICLES

R. Castañeda–Priego, A. Rodriguez–López, M. D. Carbajal–Tinoco,
P. González–Mozuelos and J. M. Méndez–Alcaraz.
*Depto. de Física, CINVESTAV-IPN, Av. IPN 2508, Col. San Pedro Zacatenco, 07360
México D. F., México.*

Abstract   We present a general method for constructing effective pair interaction
potentials in colloidal systems, which results from a contraction of the
description of colloidal mixtures based on the integral equations theory
of simple liquids. In order to illustrate its applicability, we calculate
the entropy driven potentials in mixtures of hard spheres, as well as
the screened Coulombic interactions between charged particles. The
accuracy of our results is pointed out by comparison with computer
simulations for the case of entropy driven potentials and experimental
data for the case of charged colloids.

Keywords: Liquids, interaction, colloids

1.   INTRODUCTION

Complex fluids are composed of a large amount of particles of many
different species. For example, a clean aqueous suspension of polystyrene
spheres contains, obviously, the polystyrene spheres and the water molecules, but also ions dissociated from the surface of the spheres, called
counterions, and at least two different species of salt ions. The size of
the first ones ranges from about some tens of nanometers, up to several
microns. The other particles are of subnanometer dimensions. When
looking at the suspension by means of light scattering, for example, only
the big spheres seem to be present in the systems, since they are the
only ones able to affect the photons with wavelengths of the order of
micrometers. Thus, in this case the experimental access to the system
is limited and we can only observe a part of it. The physics of this part,
however, is determined by all the existing particles, observed or not. Regardless of the kind of systems that we might be able to imagine, and of
the experimental methods that we might propose in order to look inside

Developments in Mathematical and Experimental Physics,
Volume B: Statistical Physics and Beyond
Edited by Macias *et al.*, Kluwer Academic/Plenum Publishers, 2003

3

them, the conclusion is always the same: we can observe only a part of a given system, but its physics is determined also by the rest. Physicists deal with the restricted nature of the experimental observations by developing contracted pictures, or effective theories, containing parameters related to the unobserved variables. The basic concept behind this approach is the effective pair interaction potential between the observed particles.

In this paper we present a general method for the derivation of effective pair interaction potentials between colloidal particles, which results from the contraction of the description of a multicomponent liquid. The starting point in this approach is given by the set of integral equations that describes the two–body correlations in the system. In order to illustrate its applicability, we calculate the entropy driven potentials in mixtures of hard spheres, as well as the screened Coulombic interactions between charged particles. The accuracy of our results is tested by their comparison with computer simulations and experimental data.

2.   CONTRACTION OF THE DESCRIPTION AND EFFECTIVE POTENTIALS

In an homogeneous colloidal mixture of p spherical species, the total correlation $h_{\alpha\beta}(r)$ between a particle of species α and a particle of species β, separated by the distance r, can be expressed in a logical, almost graphical form, by the expansion

$$
\begin{aligned}
h_{\alpha\beta}(r) \;=\;& c_{\alpha\beta}(r) + \sum_{\gamma=1}^{p} n_\gamma \int_V c_{\alpha\gamma}(r') c_{\gamma\beta}(|\mathbf{r}-\mathbf{r}'|)\,d\mathbf{r}' \\
& + \sum_{\gamma,\delta=1}^{p} n_\gamma n_\delta \int_V \int_V c_{\alpha\gamma}(r') c_{\gamma\delta}(r'') c_{\delta\beta}(|\mathbf{r}-\mathbf{r}'-\mathbf{r}''|)\,d\mathbf{r}'d\mathbf{r}'' \\
& + \sum_{\gamma,\delta,\epsilon=1}^{p} n_\gamma n_\delta n_\epsilon \int_V \int_V \int_V c_{\alpha\gamma}(r') \times \\
& \quad c_{\gamma\delta}(r'') c_{\delta\epsilon}(r''') c_{\epsilon\beta}(|\mathbf{r}-\mathbf{r}'-\mathbf{r}''-\mathbf{r}'''|)\,d\mathbf{r}'d\mathbf{r}''d\mathbf{r}''' \\
& + \cdots,
\end{aligned}
\tag{1}
$$

where the function $c_{\alpha\beta}(r)$ represents the 'direct' correlation between these two particles, so that $h_{\alpha\beta}(r)$ includes all possible connections between the $N = \sum_{\alpha=1}^{p} N_\alpha$ particles in the volume V; $n_\alpha \equiv N_\alpha/V$. For example, the second term of the right hand represents the links between two particles mediated by a third one, and so forth.

Equation (1) is a geometrical series and can therefore be rewritten as

$$h_{\alpha\beta}(r) = c_{\alpha\beta}(r) + \sum_{\gamma=1}^{p} n_\gamma \int_V c_{\alpha\gamma}(r')h_{\gamma\beta}(|\mathbf{r} - \mathbf{r}'|)d\mathbf{r}'. \tag{2}$$

This is the well–known Ornstein–Zernike equation [1], which has a very important property: it does not change its form when only a part of the system can be observed, i.e., it is invariant under contractions of the description. If, for example, one can only see the particles of species α, this equation can be exactly rewritten as

$$h_{\alpha\alpha}(r) = c_{\alpha\alpha}^{eff}(r) + n_\alpha \int_V c_{\alpha\alpha}^{eff}(r')h_{\alpha\alpha}(|\mathbf{r} - \mathbf{r}'|)d\mathbf{r}', \tag{3}$$

with

$$\begin{aligned}
\tilde{c}_{\alpha\alpha}^{eff}(q) &= \tilde{c}_{\alpha\alpha}(q) + \sum_{\beta=1(\neq\alpha)}^{p} \frac{n_\beta \tilde{c}_{\alpha\beta}(q)\tilde{c}_{\beta\alpha}(q)}{[1 - n_\beta \tilde{c}_{\beta\beta}(q)]} \\
&+ \sum_{\beta=1(\neq\alpha)}^{p} \sum_{\gamma=1(\neq\alpha,\beta)}^{p} \frac{n_\beta n_\gamma \tilde{c}_{\alpha\beta}(q)\tilde{c}_{\beta\gamma}(q)\tilde{c}_{\gamma\alpha}(q)}{[1 - n_\beta \tilde{c}_{\beta\beta}(q)][1 - n_\gamma \tilde{c}_{\gamma\gamma}(q)]} \\
&+ \cdots ,
\end{aligned} \tag{4}$$

where $\tilde{c}_{\alpha\beta}(q)$ is the Fourier transform of $c_{\alpha\beta}(r)$.

In a very dilute system the direct correlation between particles is given, to a very good approximation, by the negative of the pair interaction potential, $-\beta u_{\alpha\beta}(r)$, with $\beta = 1/k_B T$. This approximation is known as the mean spherical approximations (MSA) [1]. Assuming that this is also valid for concentrate systems, we get an approximated form for the effective pair interaction potentials, which, when used in the previous example, leads to the result

$$\beta u_{\alpha\alpha}^{eff}(r) = -c_{\alpha\alpha}^{eff}(r). \tag{5}$$

Therefore, the effective interactions can be calculated as a function of the unobserved variables, since $c_{\alpha\alpha}^{eff}(r)$ is given in terms of $c_{\alpha\alpha}(r)$, $c_{\alpha\beta}(r)$, $c_{\beta\beta}(r)$, etc. We illustrate this in the next section for a binary mixture of hard spheres, where our theoretical scheme leads to the entropy driven depletion forces [2].

3.   ENTROPIC FORCES

Let us suppose a binary mixture of hard spheres of diameters σ_1 and σ_2 ($\sigma_1 > \sigma_2$), and volume fractions $\varphi_1 = \pi n_1 \sigma_1^3/6$ and $\varphi_2 = \pi n_2 \sigma_2^3/6$, respectively. Moreover, we assume ourselves unable to observe species 2. Then, the effective interaction potential between particles of species 1 is $\beta u_{11}^{eff}(r) = +\infty$ for $r < \sigma_1$, and

$$\beta u_{11}^{eff}(r) = -c_{11}(r) - n_2 \mathcal{F}^{-1}\left[\frac{\tilde{c}_{12}(q)\tilde{c}_{21}(q)}{1 - n_2\tilde{c}_{22}(q)}\right] \tag{6}$$

for $r \geq \sigma_1$. The only way to evaluate $\beta u_{11}^{eff}(r)$ in a concentrate system is by means of the numerical solution of equation (2), together with an appropriate closure relation, like the Percus–Yevik (PY) or Rogers–Young (RY) approximations [3]. In the infinitely dilute limit of large particles ($\varphi_1 \to 0$), up to linear terms in n_2, however, we can get analytical expressions for the effective potential. In this case, equation (6) becomes

$$\beta u_{11}^{eff}(r) = -n_2 \mathcal{F}^{-1}\left[\tilde{c}_{12}^{(0)}(q)\tilde{c}_{21}^{(0)}(q)\right] \tag{7}$$

or

$$\beta u_{11}^{eff}(r) = -n_2 \int_V c_{12}^{(0)}(r')c_{21}^{(0)}(|\mathbf{r} - \mathbf{r}'|)d\mathbf{r}', \tag{8}$$

where $c_{\alpha\beta}^{(0)}(r) = -1$ for $r < \sigma_{\alpha\beta} = (\sigma_\alpha + \sigma_\beta)/2$, and 0 for $r \geq \sigma_{\alpha\beta}$. Thus, $c_{\alpha\beta}^{(0)}(r)$ is the first term of the density expansion $c_{\alpha\beta}(r) = c_{\alpha\beta}^{(0)}(r) + c_{\alpha\beta}^{(1)}(r)n_2 + c_{\alpha\beta}^{(2)}(r)n_2^2 + \cdots$ [1]. The symbol $\mathcal{F}^{-1}$ stands for the inverse Fourier transform.

The integral in equation (8) clearly accounts for the volume of the region from which the particles of species 2 are excluded, V^{exc}, when the two particles of species 1 involved are separated by the distance r. This equation can therefore be written as $u_{11}^{eff}(r) = -(k_B T N_2/V)V^{exc}$. Since the term in parenthesis corresponds to the pressure of an ideal gas composed of particles of species 2, the effective interaction potential is equivalent to the increment of the free energy of the ideal gas when its volume is reduced by V^{exc}. This is the theory of Asakura–Oosawa (AO) for depletion forces [4], i.e., for the forces resulting from the expulsion of small particles from the gap between two approaching large particles. This becomes evident by evaluating equation (7) or (8), since in this case we reproduce the result of AO for binary mixtures of hard spheres [2, 4]:

$$\beta u_{11}^{eff}(r) = -\varphi_2\left[(\eta + 1)^3 - \frac{3}{2}(\eta + 1)^2\frac{r}{\sigma_2} + \frac{1}{2}\frac{r^3}{\sigma_2^3}\right] \tag{9}$$

for $\sigma_1 \leq r \leq \sigma_1 + \sigma_2$ and 0 for larger distances, with $\eta = \sigma_1/\sigma_2$.

Our theoretical scheme not only contains the AO theory for depletion forces (in the limit of infinite dilution), but also the general expressions for concentrate systems. In order to check the accuracy of the latter, we compare its results with molecular dynamics (MD) simulations for a bidimensional (2D) binary mixture of hard disks of diameters σ_1 and σ_2 ($\sigma_1 > \sigma_2$), and surface fractions $\varphi_1 \to 0$ and $\varphi_2 = \pi n_2 \sigma_2^2 / 4$. The simulated depletion forces were obtained by summing the linear moment exchange over all collisions between all the particles of species 2 and two fixed particles of species 1, separated by the distance r. The depletion potentials result from the integration of the forces [5]. The AO limit is obtained by evaluating equation (7) or (8) in 2D, where $\mathcal{F}^{-1}$ stands for the inverse Fourier–Bessel transform and dr for a differential of area [6]:

$$
\beta u_{11}^{eff}(r) = -\frac{2\varphi_2}{\pi}(\eta+1)^2 \left[\arccos\left(\frac{1}{1+\eta}\frac{r}{\sigma_2}\right) \right.
$$
$$
\left. -\frac{1}{1+\eta}\frac{r}{\sigma_2}\sqrt{1 - \left(\frac{1}{1+\eta}\frac{r}{\sigma_2}\right)^2} \right] \tag{10}
$$

for $\sigma_1 \leq r \leq \sigma_1 + \sigma_2$ and 0 for larger distances. The effective interaction potential is also obtained by numerically solving the OZ equation together with the PY closure relation [6]. Once the correlations have been calculated, the resulting direct correlation functions $c_{11}(r)$, $c_{12}(r) = c_{21}(r)$ and $c_{22}(r)$ are used as input to equation (6) for the evaluation of $\beta u_{11}^{eff}(r)$.

The depletion potentials are shown in Figure 1 for $\varphi_2 = 0.21$ and 0.42, with a size ratio $\eta = 5$. The symbols correspond to the simulations (MD), the dashed lines to the AO theory, and the full lines to the PY approximation. The attraction at contact becomes deeper with increasing concentration of small particles, which can be expected from the AO limit. The simultaneous development of a potential barrier in front of this attractive well is really surprising, as well as the fact that the interaction becomes more and more long-ranged. This behavior is due to correlations between small particles, which are not included in the AO model, but they are taken into account in the PY approach. Our theory captures all concentration effects with quantitative accuracy.

4.   INTERACTION BETWEEN CHARGED PARTICLES

The handling of effective interactions between charged particles requires some additional considerations. On one hand, the already studied entropic interactions are still present. On the other hand, Coulombic systems are characterized by an interaction term of the form $l_B z_\alpha z_\beta / r$, where $l_B = \beta e^2 / \epsilon$ is Bjerrum's length, e and ϵ are the electron charge

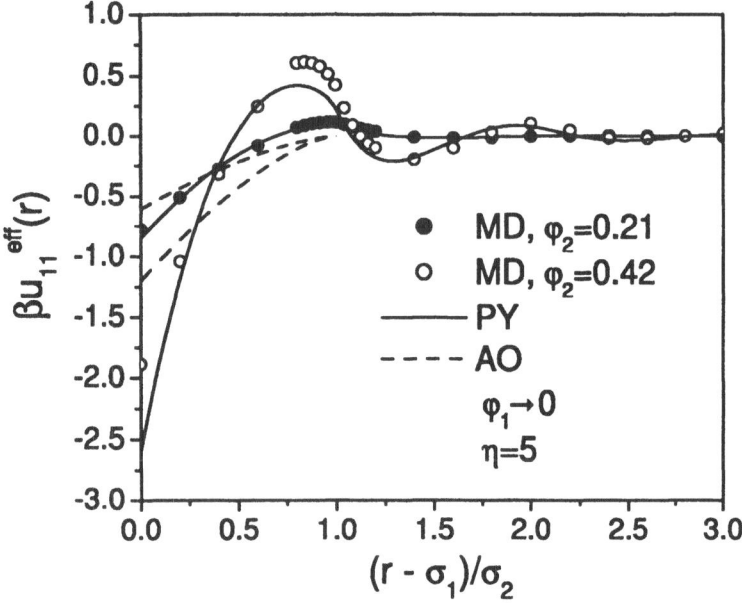

Figure 1. The figure shows the depletion potential $\beta u_{11}^{eff}(r)$ for bidimensional binary mixtures of hard disks with diameters $\sigma_1 = 5\sigma_2$ and surface fractions $\varphi_1 \to 0$ and $\varphi_2 = 0.21$ and 0.42. The symbols correspond to the simulations (Molecular Dynamics), the dashed lines to the AO theory, and the full lines to the PY approximation.

and the solvent dielectric constant, respectively (for water at room temperature, $l_B \approx 7$ Å), and z_α is the valence of species α. The long ranged Coulombic term can be related to the asymptotic part of direct correlation functions in the following way,

$$c_{\alpha\beta}(r) = c_{\alpha\beta}^s(r) - \frac{l_B}{r} z_\alpha z_\alpha, \qquad (11)$$

where $c_{\alpha\beta}^s(r)$ is the short ranged part of the direct correlation function $c_{\alpha\beta}(r)$. The formal solution of the Ornstein–Zernike equation for Coulombic systems can be obtained by Fourier transforming equations (2) and (11). The result is expressed in terms of the short ranged direct correlations functions, and, as shown in ref. [7], this formal solution directly satisfies the Stillinger–Lovett moment conditions [8]. Moreover, an effective pair potential between charged macroparticles can be defined in terms of the computed direct correlation functions [7]. This effective potential was compared to experimental measurements performed in the dilute limit [9]. As expected, the calculated and the measured effective potentials showed a Yukawa–like behavior. However, when the macroion volume fraction is increased, while keeping fixed the other parameters of the system, we get as a result strong deviations from the screened Coulomb form.

Our specific model for the phenomena described in the present section is based on the primitive model, and it consists of three species of ions. The hard–core macroions (species 1) have valence $z_1 = -Z$ ($Z \gg 1$) and diameter $\sigma_1 = \sigma$. The small ions (species 2 and 3) are modelled as point ions. Thus, $c_{\alpha\beta}^s(r) = 0$ for $\alpha, \beta = 2, 3$. This approximation seems reasonable, provided that there is a large asymmetry in size between the particles used in the experiments [9] and the small ions of the electrolyte. The remaining valences are $z_2 = 1$ (counterions) and $z_3 = -1$ (coions). For this specific model, the total correlation functions between macroions and macroions–small ions are given in Fourier space by the expressions [7],

$$\tilde{h}_{\alpha\beta}(q) = \tilde{h}_{\alpha\beta}^s(q) - \frac{4\pi l_B}{q^2 + \kappa_T^2(q)} \tilde{z}_\alpha^*(q)\tilde{z}_\beta^*(q), \tag{12}$$

for $\alpha, \beta = 1, 2, 3$, where

$$\tilde{h}_{11}^s(q) = \frac{\varphi(q)}{1 - n_1\varphi(q)}, \tag{13}$$

$$\tilde{h}_{1\gamma}^s(q) = \frac{\tilde{c}_{1\gamma}^s(q)}{1 - n_1\varphi(q)}, \tag{14}$$

$$\tilde{z}_1^*(q) = \frac{-Z + n_2\tilde{c}_{12}^s(q) - n_3\tilde{c}_{13}^s(q)}{1 - n_1\varphi(q)}, \tag{15}$$

$$\tilde{z}_\gamma^*(q) = z_\gamma + n_1\tilde{c}_{1\gamma}^s(q)\tilde{z}_1^*(q), \tag{16}$$

$$\kappa_T^2(q) = 4\pi l_B \sum_{\delta=1}^{3} z_\delta n_\delta \tilde{z}_\delta^*(q), \tag{17}$$

for $\gamma, \delta = 2, 3$, and

$$\varphi(q) = \tilde{c}_{11}^s(q) + n_2(\tilde{c}_{12}^s(q))^2 + n_3(\tilde{c}_{13}^s(q))^2 \tag{18}$$

The above equations must be complemented with a set of closure relations. Let us mention that, in an early attempt, Medina–Noyola and McQuarrie [10] retrieved the effective pair potential of Derjaguin, Landau, Verwey and Overbeek (DLVO) [11], for the case of highly dilute macroion concentrations. These authors used the MSA closure for all correlations. Within this approach, however, some unphysical results can be obtained. That is the case of a negative radial distribution function between macroparticles. The HNC approximation, on the other hand, produces a considerable overestimation of the attractive correlations.

The strong coupling present in highly charged colloidal systems suggests the use of a more robust approach. Promising results have been obtained with the so–called self–consistent closure approximations. For example, the RY approximation is a phenomenological mixture of the PY and HNC closures and it has been used successfully in the description of purely repulsive interactions. In order to treat attractive interactions, Zerah and Hansen (ZH) proposed an additional approximation in which the MSA and the HNC closures are mixed in a similar way as in the RY case [12]. In the ZH approach, the correlations involving macroions are

$$g_{12}(r) = \theta(r - \sigma_{12}) \left[1 + \frac{1}{f_{12}(r)} \left(\exp\left[f_{12}(r) \left(\gamma_{12}(r) - Z\frac{l_B}{r} \right) \right] - 1 \right) \right],$$
(19)

and

$$g_{1\alpha}(r) = \theta(r - \sigma_{1\alpha}) \exp\left(-|z_\alpha| Z\frac{l_B}{r} \right) \times$$
$$\left[1 + \frac{1}{f_{1\alpha}(r)} \left(\exp\left[f_{1\alpha}(r)\gamma_{1\alpha}(r) \right] - 1 \right) \right],$$
(20)

for $\alpha = 1, 3$. Here $\gamma_{\alpha\beta}(r) \equiv h_{\alpha\beta}(r) - c_{\alpha\beta}(r)$, $\theta(r)$ is the step function that takes into account the hard core, and $f_{\alpha\beta}(r) \equiv 1 - \exp(-\alpha_{\alpha\beta}r)$ is the corresponding mixing function with its corresponding mixing parameter $\alpha_{\alpha\beta}$. The adjustable parameters $\alpha_{1\beta}$ are determined through the following partial thermodynamic consistency criterion. The partial pressure that comes from the contribution of macroparticles is defined by considering a semipermeable membrane that only allows the transit of small ions through it, i.e.,

$$\frac{\beta P_{v1}}{n_1} = 1 - \frac{2\pi}{3} \sum_{\beta=1}^{3} n_\beta \int_0^\infty dr\, r^3 \frac{d\beta u_{1\beta}(r)}{dr} g_{1\beta}(r).$$
(21)

A partial compressibility is then obtained by differentiating numerically this last equation with respect to the salt concentration while maintaining the number density of the macroparticles constant,

$$\chi_{v1s}^{-1} = \left(\frac{\partial \beta P_{v1}}{\partial n_s} \right)_{T,n_1},$$
(22)

where $n_s = n_2 + n_3$. A second route to calculate this partial compressibility is assumed here to be given by the equation [1]

$$\chi_{c1s}^{-1} = 1 - 4\pi \frac{n_1}{n} \sum_{\beta=1}^{3} n_\beta \int_0^\infty dr\, r^2 \, c_{1\beta}(r),$$
(23)

where $n = n_1 + n_2 + n_3$ is the total number density. The numerical values of χ_{c1s}^{-1} and χ_{v1s}^{-1} certainly depend on the mixing parameters. By requiring $\chi_{c1s}^{-1} = \chi_{v1s}^{-1}$ we get at least a partial thermodynamic consistency. This condition, however, does not define a unique set of mixing parameters $\alpha_{1\beta}$. According to a previous work [7], the appropriate mixing parameters are those that minimize the grand potential $\Omega_1 = -P_1 V_1$ with the restriction $\chi_{c1s}^{-1} = \chi_{v1s}^{-1}$, where P_1 is extracted from equation (21) and V_1 is the region in which the macroparticles are confined.

In all the cases studied here, the ZH closure becomes practically identical to the HNC closure for the correlations between macroions, $h_{11}(r)$, thus, the Fourier transform of the effective pair potential between charged macroparticles takes the form [7],

$$\beta \tilde{u}_{11}^{eff}(q) = \beta \tilde{u}_{11}^{HS}(q) + \frac{4\pi l_B}{q^2 + \kappa_s^2} \tilde{z}_1^r(q)^2 - n_2 (\tilde{c}_{12}^s(q))^2 - n_3 (\tilde{c}_{13}^s(q))^2, \quad (24)$$

where $\beta \tilde{u}_{11}^{HS}(k)$ is the hard sphere term (such that $\beta u_{11}^{HS}(r) = 0$ for $r > \sigma$), and

$$\tilde{z}_1^r(q) = -Z + n_2 \tilde{c}_{12}^s(q) - n_3 \tilde{c}_{13}^s(q) \quad (25)$$

is the renormalized macroparticle charge distribution.

The input parameters used in our model were extracted from a fit to the experimental data of Crocker and Grier [9]. These measurements are especially interesting because the interaction pair potential was determined in the limit of very low macroparticle concentration. The parameter values determined, which we use here as fixed input values, are: $\sigma = 0.65$ μm, $Z = 4694$, and $\kappa_s^{-1} = 272$ nm; where

$$\kappa_s = \sqrt{4\pi l_B (n_2 z_2^2 + n_3 z_3^2)} \quad (26)$$

is the inverse screening length due to the small ions. The volume fraction of macroions, $\varphi = \pi n_1 \sigma^3 / 6$, is assumed to be 10^{-6}, according to the low concentration regime of the experiments. Here we analyze the effect of increasing the volume fraction φ, while keeping fixed the screening length.

In Figure 2 we present the effective pair potential obtained from equation (24) for three different volume fractions, namely, 10^{-6}, 0.00348, and 0.00395. The squares represent the experimental data of Crocker and Grier [9]. As soon as the volume fraction is increased, the effective pair potential shows a progressive overscreening until the appearance of an attractive component. The potential of curve in full line has a depth of about 0.5 $k_B T$. Although the prediction of such curve has not being compared with experimental equilibrium data, it is consistent with the

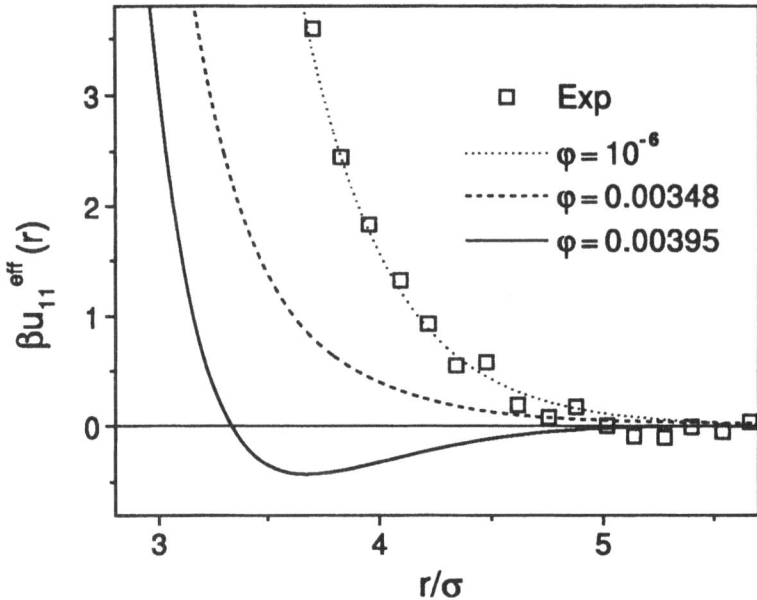

Figure 2. The figure shows the effective pair potential between macroparticles corresponding to the values $\sigma = 0.65$ μm, $Z = 4694$, $\kappa_s^{-1} = 272$ nm, for the following macroion volume fractions: $\varphi = 10^{-6}$ (dotted line), $\varphi = 0.00348$ (dashed line), and $\varphi = 0.00395$ (full line). The squares represent the experimental data from Ref. 9.

observations done on metastable crystallites by Larsen and Grier in a quite similar system [13]. Such crystallites were formed in an electrolyte of basically constant screening length. Moreover, the separation between nearest neighbors in the crystallite is of about 3σ, which is comparable to the position of the well of our model. Additionally, the decrement of the repulsion observed in the curves of Fig. 2 is in clear qualitative agreement with the experimental observations recently reported by Behrens and Grier for a quasi–bidimensional colloidal system [14].

5.   DISCUSSION AND CONCLUSIONS

We presented here a general method for the derivation of effective pair interaction potentials between colloidal particles, which results from a contraction of the description of multicomponent systems based on the integral equations theory of simple liquids. Its applicability was illustrated by calculating the entropy driven potentials in mixtures of hard spheres, as well as the screened Coulombic interactions between charged particles. The comparison of our results with computer simulations of depletion forces confirms that our theoretical scheme goes beyond the Asakura–Oosawa theory (dilute limit of our equations) and captures all concentration effects with quantitative accuracy. In particular, the increment of the attraction at contact, the simultaneous development of a repulsive barrier in front of this, and the increasing range of the interaction. In the case of charged colloids, our approach allowed us to go beyond the Yukawa model.

Although not shown in this paper, our theoretical scheme can also be used in order to calculate dielectric functions in charged systems [15], as well as in the design of surfaces of entropic potential for the selective adsorption of colloidal particles. These results will be published elsewhere.

Acknowledgments

The authors thank CONACyT–Mexico for financial support (Grants 33815–E and 36464–E).

References

[1] J. P. Hansen and I. R. McDonald, *Theory of Simple Liquids* (Academic Press, London, 1986).

[2] J. M. Méndez–Alcaraz and R. Klein, Phys. Rev. **E61** (2000) 4095.

[3] J. M. Méndez–Alcaraz, B. D'Aguanno and R. Klein, Langmuir **8** (1992) 2913.

[4] S. Asakura and F. Oosawa, J. Chem. Phys. **22** (1954) 1255.

[5] A. Rodríguez–López, Master's degree Thesis, CINVESTAV–IPN (2002).

[6] R. Castañeda–Priego, Master's degree Thesis, CINVESTAV–IPN (2001).

[7] P. González–Mozuelos and M. D. Carbajal–Tinoco, J. Chem. Phys. **109** (1998) 11074.

[8] F. H. Stillinger and R. Lovett, J. Chem. Phys. **49** (1968) 1991.

[9] J. C. Crocker and D. G. Grier, Phys. Rev. Lett. **77** (1996) 1897.

[10] M. Medina-Noyola and D. A. McQuarrie, J. Chem. Phys. **73** (1980) 6279.

[11] B. V. Derjaguin and L. Landau, Acta Physicochim. URSS **14** (1941) 633.

[12] G. Zerah and J.-P. Hansen, J. Chem. Phys. **84** (1986) 2336.

[13] A. E. Larsen and D. G. Grier, Nature (London) **385** (1997) 230.

[14] S. H. Behrens and D. G. Grier, Phys. Rev. **E64** (2001) 050401.

[15] P. González–Mozuelos and N. Bagatella–Flores, Physica **A286** (2000) 56.

SELF–DIFFUSION OF COLLOIDAL PARTICLES IN TWO–DIMENSIONAL POROUS MEDIA

Angeles Ramirez–Saito and José Luis Arauz–Lara
Instituto de Física "Manuel Sandoval Vallarta"
Universidad Autónoma de San Luis Potosí, Alvaro Obregón 64, 78000 San Luis Potosí, SLP, México
saito@ifisica.uaslp.mx, arauz@ifisica.uaslp.mx

Abstract   Here we report a study of the short–time motion of colloidal particles in two–dimensional porous media. A single layer of colloidal particles is fixed between two glass plates forming a disordered array of obstacles. The void space is filled with an aqueous suspension of colloidal particles of smaller size. The lateral diffusion of the smaller species is measured in real space by optical microscopy. Quantities describing single particle motion, such as the step distribution functions and the particles mean squared displacement, are measured in the presence and absence of obstacle particles. At very short times, the motion of individual particles is qualitatively insensitive to the presence of the obstacles. However, at intermediate times the motion of individual particles is quite sensitive to the local distribution of obstacles. The measured quantities, when averaged over the trajectories of particles diffusing in different pores, show a behavior corresponding to particles diffusing in a homogeneous medium with a lower effective mobility.

Keywords: Brownian motion, porous media, colloidal suspension

1.   INTRODUCTION

An understanding of the motion of colloidal particles, such as polymers, proteins, DNA, aggregates, through a disordered porous medium, or through a matrix of slow dynamics, is relevant in many fields such as oil recovery, chromatography, catalysis dialysis, etc. [1, 2, 3, 4, 5]. It is clear that the physical properties of a colloidal fluid in a porous matrix depend on quantities such as the matrix structure and the interactions (direct and hydrodynamic) of the colloidal particles between themselves and with the matrix. Dynamical properties would also depend on the

Developments in Mathematical and Experimental Physics,
Volume B: Statistical Physics and Beyond
Edited by Macias *et al.*, Kluwer Academic/Plenum Publishers, 2003

time and space scale considered. For homogeneous systems, a distinction between short– and long–time behavior is introduced by the effect of the direct interactions between the particles [6]. In the case of diffusion in a porous matrix, there is also a time scale separation, even for single particles, imposed now by the interactions with the matrix. For instance, for isolated particles in the bulk, the mean squared displacement $W(t) \equiv \langle [\Delta \mathbf{r}(t)]^2 \rangle /6$ is a simple linear function of time, i. e., $W(t) = D_0 t$ with D_0 being the free diffusion coefficient. However, in a porous media a single particle might undergo anomalous diffusion, i. e., $W(t) \sim t^\lambda$ with $\lambda \neq 1$, close to the percolation limit of the solid phase. This behavior should be best observed at short times where a particle diffusing inside a cavity probes the inhomogeneous local environment. In the long–time regime, on the other hand, a particle diffuses through different pores probing now the characteristic structure of the matrix. In this case the motion of the particle can be characterized by an effective diffusion constant D_L, which might undergo a critical behavior as the volume fraction ϕ of the solid phase of the matrix increases and approaches its percolation limit value Φ, i. e., $D_L \sim (\Phi - \phi)^\delta$, where δ is the critical exponent [7, 8]. At finite concentrations of diffusing particles, the interactions between the particles and between them and the matrix increase the complexity of these phenomena. In this case the description of the dynamic properties is also more complex since any theoretical approach should incorporate these interactions explicitly.

In this work, we focus on a experimental study of the short–time dynamics of colloidal particles in a two–dimensional (2D) array of obstacles using optical techniques. The system consist of a binary mixture of colloidal particles in water at room temperature ($24\,^\circ C$). One of the species (species 1) being fluorescent polystyrene spheres of diameter σ_1, and the other species (species 2) are non–fluorescent polystyrene spheres of diameter $\sigma_2 > \sigma_1$. The suspension is confined between two glass plates which are pressed against each other until the motion of the larger particles is completely quenched. Thus, the particles of species 2 are frozen in a 2D disordered configuration forming an effective 2D porous matrix. Then, the mobile particles of species 1 constitutes a colloidal fluid which equilibrates in the static field of the fixed particles. In our experiment both species are in the micrometer size range and they can be imaged using an optical microscope. Thus, the motion of the mobile species and the structure of the the matrix can be accurately determined using video–microscopy. The experimental details on sample preparation and data analysis are given in section 1, and the results are presented and discussed in section 2.

2. METHODS

2.1. SYSTEM PREPARATION

Aqueous suspensions of fluorescent polystyrene spheres of diameter $\sigma_1 = 1.01 \pm 3\%$ μm and non–fluorescent polystyrene spheres of diameter $\sigma_2 = 2.02 \pm 3\%$ μm (Duke Scientific), were extensively dialyzed (dialysis bag of 50,000 molecular weight cut-off) against nanopure water to eliminate the surfactants from the original batches. In a clean atmosphere of nitrogen gas, the suspension of small particles is mixed with the suspension of large particles. A little volume of the mixture (≈ 1 μl) is confined between two carefully cleaned glass plates (a slide and a cover slip), which are uniformly pressed one against the other until the separation h between the plates coincides with σ_2. As a result, the large particles are fixed in a disordered configuration across the sample, serving as obstacle particles to the smaller species and at the same time as spacers between the plates. The system is then sealed with epoxy resin (Epo-Tek 302), and the species of small mobile particles allowed to equilibrate in this confined geometry for few days at room temperature (24 ± 0.1 $°C$). Under these conditions, the motion of the diffusing species along the direction z, perpendicular to the plates, is restricted to a small fraction of σ_1. Although the systems is not strictly 2D, the main motion of the particles is along the plane (x,y) parallel to the walls. Thus, we study only the lateral motion of the particles in the matrix of obstacles. Figure 1 shows an image of area $a = 37 \times 28$ μm^2 of a sample with mobile and fixed particles area fractions $\phi_1 = 1 \times 10^{-2}$ and $\phi_2 = 1.3 \times 10^{-1}$, respectively, where $\phi_i \equiv \pi n_i^*/4$, with $n_i^* \equiv n_i \sigma_i^2$ being the reduced concentration of particles of species i and n_i the average number of particles of that species in the area a. The image was taken using an optical microscope with a 100× objective.

2.2. DIGITAL VIDEO MICROSCOPY

As mentioned above, the systems are observed in real space using an optical microscope with a 100× objective. The motion of the particles is recorded using a CCD camera coupled to a video tape recorder. Images are digitized (see Fig. 1) using a frame grabber with a resolution of 640×480 pixels. With our setup we measure $\sigma_1 = 17.54$ pixels. The position (i. e., the x and y coordinates) of every particle in the field of view is determined from the digitized images using the method devised by Crocker and Grier [9], which allows to locate the spheres' centers with a precision of 1/5 pixel ($\sim 0.01\sigma_1$). The particles' motion in our system is slow, they move in average only a fraction of σ_1 between frames.

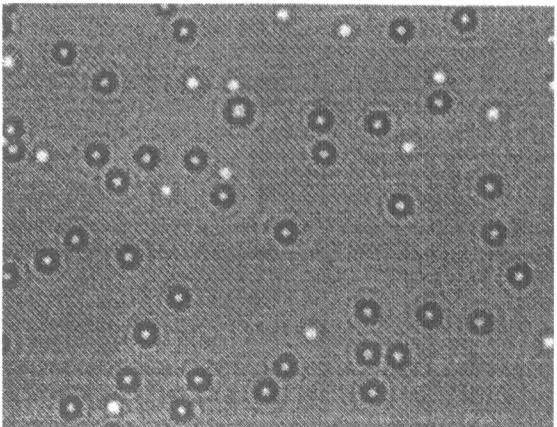

Figure 1.   Optical image, area $a = 37 \times 28 \ \mu\mathrm{m}^2$, of a typical configuration of 2.02 μm polystyrene spheres fixed between two glass plates (large dark circles with bright spot in the center) and 1.01 μm mobile polystyrene spheres (bright circles). The picture was taken using an optical microscope with a 100× objective

Thus, their 2D trajectories can be easily reconstructed, with a time resolution of 1/30 s, from their positions at consecutive frames. From the trajectories various physical quantities of interest can be obtained as we explain below.

3.   RESULTS

We studied the motion of the mobile species in matrices of fix particles at different concentrations far from the percolation limit of the solid phase. Since we found similar qualitative behavior of the mobile particles for different values of ϕ_2, here we present only two illustrative cases: $\phi_2 \sim 0$ and $\phi_2 = 0.18$. In what follows, the mobile particles will be referred to as the particles whereas the fix particles will be referred to as the obstacles or as the matrix. Since we are mainly interested in the effects of the matrix on the particles motion, we study systems with a very low concentration of the mobile species ($\phi_1 \sim 1 \times 10^{-2}$) in order to minimize the effects of the interparticle interactions.

The simplest quantity describing the lateral motion of individual particles is the mean squared displacement,

$$W(t) = \frac{1}{4}\langle [\mathbf{r}(t) - \mathbf{r}(0)]^2 \rangle, \qquad (1)$$

where $\mathbf{r}(t)$ is the 2D particle's position at time t and the angular brackets represent an equilibrium ensemble average. This quantity describes single particle motion and should be sensitive to the effects of the local envi-

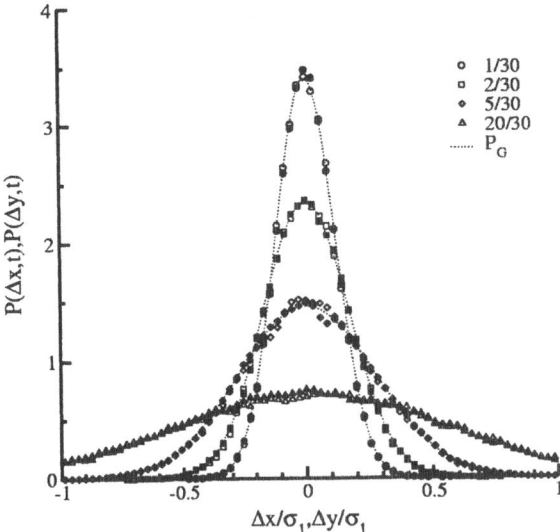

Figure 2. Measured distribution functions $P(\Delta x, t)$ and $P(\Delta y, t)$ of steps, Δx and Δy, in the directions x and y, respectively, at different times (symbols). For this system $\phi_1 = 0.016$ and $\phi_2 \sim 0$. The dotted lines correspond to Gaussian functions $P_G(\Delta x, t)$ constructed with the measured $W(t)$ (see text).

ronment. In the case of very low concentration of obstacles ($\phi_2 \approx 0$), the 2D displacement of the particles is expected to be random and isotropic with a (nearly) Gaussian distribution [10, 11]. Let us consider first the simple case where no obstacle particles are present ($\phi_2 \sim 0$) and the concentration of mobile particles is finite but low, $\phi_1 = 0.016$. Fig. 2 shows the distribution functions $P(\Delta x, t)$ and $P(\Delta y, t)$ of steps, Δx and Δy, in the directions x and y, respectively, for different times (symbols). As one can see in this figure, the distributions of steps along the perpendicular directions x and y are equivalent. They are symmetric functions, centered around $\Delta x, \Delta y = 0$, initially very narrow and then for larger times they spread out due to self–diffusion of the particles. One can also see that both sets of functions are very well approximated by Gaussian functions. In this figure the dotted lines correspond to Gaussian functions $P_G(\Delta x, t)$ having zero mean value and dispersion $\sqrt{2W(t)}$ [12], i. e.,

$$P_G(\Delta x, t) = \frac{1}{\sqrt{4\pi W(t)}} \exp\left[-\frac{\Delta x^2}{4W(t)}\right]. \qquad (2)$$

Thus, in this case (vanishing concentration of obstacles) the system is essentially an affective 2D system, homogeneous along the plane parallel

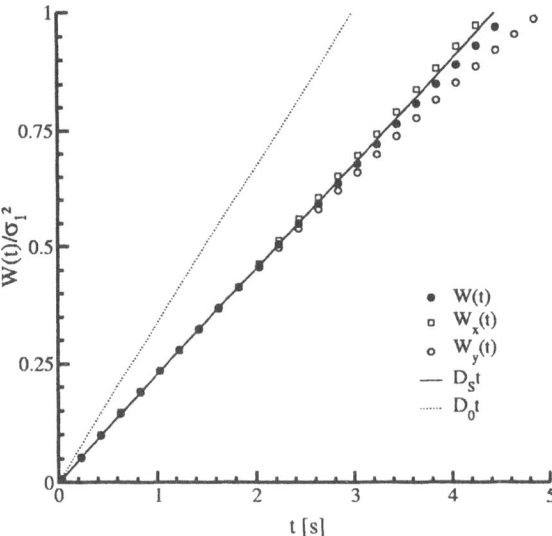

Figure 3.   Mean squared displacement of the particles in the system of figure 1 (closed circles). The open symbols correspond to the mean squared displacements $W_x(t)$ and $W_y(t)$, measured along the directions x and y, respectively. The solid line represents the initial slope of $W(t)$ and the dotted line corresponds to free diffusion D_0t.

to the plates. Figure 3 shows the measured mean squared displacement $W(t)$ (closed circles) for the particles in this system as defined in Eq. 1. We also present here (open symbols) the mean squared displacement measured along the perpendicular directions x and y, $W_x(t) \equiv \langle[x(t) - x(0)]^2\rangle/2$ and $W_y(t) \equiv \langle[y(t) - y(0)]^2\rangle/2$, respectively. Here again, one can see that the diffusion of the particles is quite similar along both perpendicular directions. Since both direction are equivalent we can write $2W(t) = \langle[\Delta x(t)]^2\rangle = \langle[\Delta y(t)]^2\rangle$. In this figure the solid line is the initial slope D_S of $W(t)$, which was obtained by fitting the short time experimental data up to $t = 1$ s to a straight line. For comparison, we also show the mean squared displacement of a free particle of the same size diffusing in the bulk D_0t (dotted line). The quantity D_S is referred to as the short–time self–diffusion coefficient. For the system in figure 2 $D_S = 2.31 \times 10^{-9}$ cm^2/s whereas $D_0 \equiv K_BT/3\pi\eta\sigma_1 = 3.43 \times 10^{-9}$ cm^2/s, K_BT being the thermal energy and η the solvent's viscosity. As one can see here, the short–time diffusion–coefficient D_S is less that 70% of its free diffusion value D_0. Since particle–particle interactions are negligible, this reduction in the particles' mobility can only be due to the effect of particle–walls interactions which main contribution comes

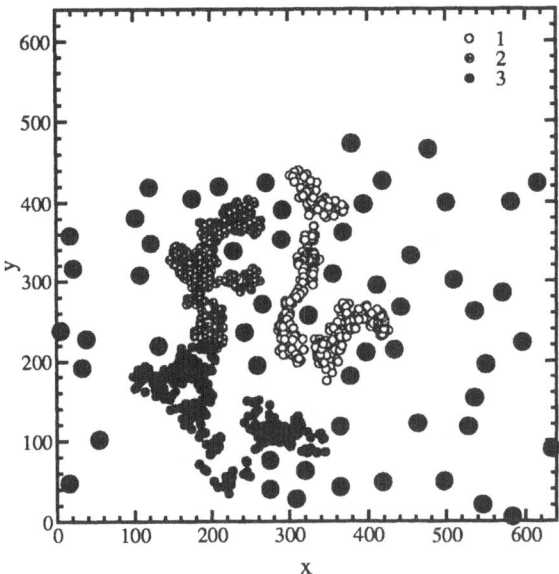

Figure 4. A typical configuration of obstacle particles (large black circles) and 3 trajectories of mobile particles (small circles). Here one can see that the particles' trajectories closely resemble the porous structure of the matrix.

from the hydrodynamic coupling of the particles motion with the walls [11].

Let us now consider the case of a finite concentration of obstacles and their effects on the dynamic properties of the mobile particles. Then, we consider here a system with $\phi_2 = 0.18$ and $\phi_1 = 0.01$, prepared as explained above, and report results for physical quantities such as the step distribution functions and the mean squared displacement of the mobile species. Figure 4 shows an actual configuration of obstacle particles in the system studied (large black circles) and the trajectories of 3 mobile particles (small circles) diffusing throughout this array for a time interval of 120 s. As one can see here, the trajectory of a single particle, tracked for a sufficient long time, closely resembles the shape of the local matrix structure. This, of course, is due to the presence of the obstacles which allow the particle to move only in the void spaces. Since the obstacles are fixed in a disordered configuration, the voids are also disordered and irregular in size and shape. Thus, the isotropy of the system is locally loss. The effects of this anisotropy can be quantified by a detailed analysis of the particles' motion. Let us analyze first only the motion of one of the particles in figure 4. Thus, we consider here only particle 1. Furthermore, let us consider only the part of the trajectory

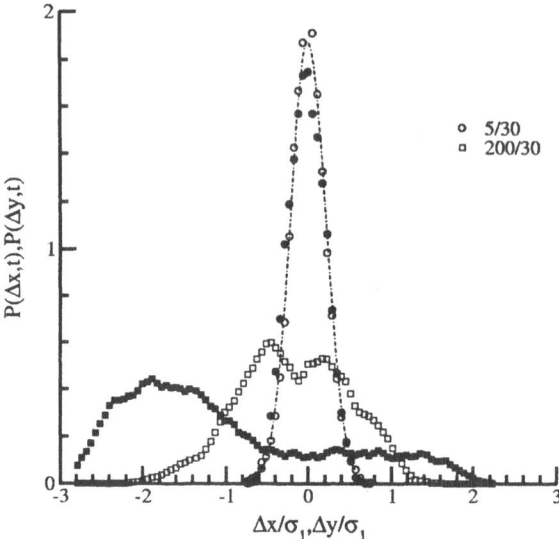

Figure 5.   Distribution functions of steps measured in the elongated part of the trajectory of particle 1 in figure 4. The open symbols represent the distributions of steps along the x direction and the closed symbols the distributions along the y axis. Circles correspond to a short time $t = 5/30$ s, and the squares to a longer time $t = 20/3$ s. The dashed–dotted line is the Gaussian function $P_G(\Delta x, t)$ constructed with $W(t)$.

where the particle moves in the void elongated along the y direction. At very short times the motion of the particles is still isotropic, i. e., the distribution of steps along the perpendicular directions are still equivalent and symmetric. This can be seen in figure 5 where $P(\Delta x, t)$ and $P(\Delta y, t)$ are shown for $t = 5/30$ s, open and close circles, respectively. These quantities are also well described by Gaussian functions $P_G(\Delta x, t)$ (dashed–dotted line). Thus, at short times the particles diffuse qualitatively as if they were in an homogeneous solvent. At larger times the behavior is different: the distributions of steps are non–Gaussian, non–symmetric and $P(\Delta x, t) \neq P(\Delta y, t)$. Figure 5 shows the distributions of steps also for $t = 20/3$ s (squares). At this time, the behavior of the distributions of steps inside the chosen void is quite different from that at short times. In the direction x the distribution (open squares) is approximately symmetric around $\Delta x = 0$ while the steps in the direction y (closed squares) are distributed preferentially towards the negative values as if the particle were flowing in that direction. Naturally, this peculiar behavior is nothing but the expression of the local anisotropy of the matrix, and different results are obtained at different voids.

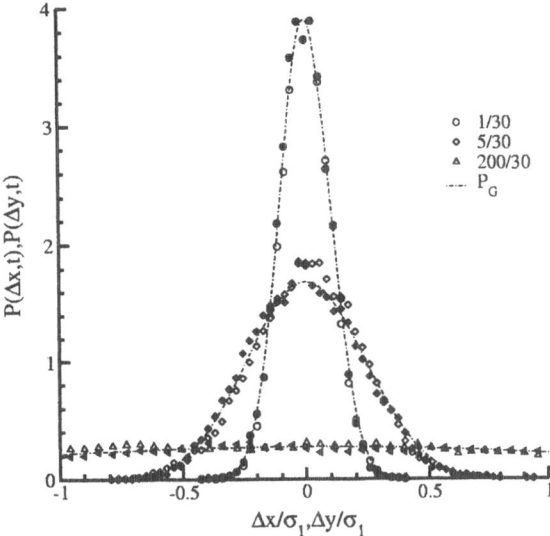

Figure 6. Distribution functions of steps averaged over the trajectories of several particles diffusing in different pores. Open and closed symbols correspond to $P(\Delta x, t)$ and $P(\Delta y, t)$, respectively. The averaged distributions are symmetric, equivalent along the different directions, and well described by the Gaussian functions $P_G(\Delta x, t)$ (dashed-dotted lines).

A global description of the dynamic properties of the colloidal particles diffusing in a matrix of obstacles as the one considered here, should take into account the particles' dynamics at different voids in the sample. Figure 6 shows the step distribution functions $P_G(\Delta x, t)$ (open symbols) and $P_G(\Delta y, t)$ (closed symbols) averaged over the trajectories of several particles at different system's locations. It is interesting to note that the averaged distribution functions recover the properties of these functions describing the diffusion of particles in an homogeneous system, i. e., they are symmetric around $\Delta x, \Delta y = 0$, $P(\Delta x, t) = P(\Delta y, t)$, and they follow the Gaussian shape of $P_G(\Delta x, t)$ (dashed–dotted line). Thus, the average motion of colloidal particles diffusing in a matrix of disordered obstacles is qualitatively similar to their motion in the absence of obstacles. There are, however, quantitative differences. The most apparent is the reduction of the particles' mobility, due to the obstacles. This can be seen in figure 7 where the averaged mean squared displacements are shown (symbols). The solid line is a straight line representing the initial slope $D_S(\phi_2 = 0.18)$ of the measured mean squared displacement $W(t)$. For comparison, the initial slope of $W(t)$ in the absence of obstacles $D_S(\phi_2 \approx 0)$ is also shown in figure 7 (dashed line). Thus, as one can

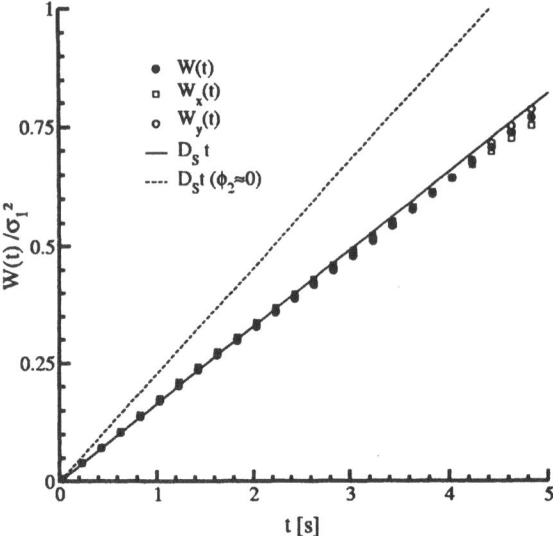

Figure 7.   Mean squared displacements averaged over several particles diffusing in different voids of the sample (symbols). The solid and dashed lines represent the initial slope $D_S(\phi_2 = 0.18)$ and $D_S(\phi_2 \approx 0)$ of the mean squared displacement. The net effect of the obstacle particles is an appreciable reduction in the effective mobility of the diffusing particles.

see here, the effect of the matrix is the net reduction of the diffusion constant of the diffusing particles.

4.   SUMMARY

In this work we investigate the diffusion properties of single (colloidal) particles in a 2D matrix of obstacles. We found that at very short–times, the motion of the particles is qualitatively similar to their motion in the absence of obstacles, i. e., quantities such as the step distribution functions or the mean squared displacements show essentially the same behavior independently of the obstacle particles concentration. At longer times, the particles probe the local shape of the matrix leading to a differentiation in the particles motion along the different directions. However, the average motion of the particles in the matrix is qualitatively similar to the motion in a homogeneous system but with a lower mobility. This qualitative similarity is likely due to the disordered distribution of obstacles. A periodic matrix of obstacles would lead to different results.

Acknowledgments

This work was partially supported by Consejo Nacional de Ciencia y Tecnología (CONACyT, México), Grants G29589E and ER026 Materiales Biomoleculares, and by Instituto Mexicano del Petróleo, Grant FIES-98-101-I.

References

[1] M. Sahimi, Phys. Rep. **306** (1998) 213.

[2] G. Viramontes–Gamboa, J. L. Arauz–Lara, and M. Medina–Noyola, Phys. Rev. Lett. **75**, 759 (1995); G. Viramontes–Gamboa, M. Medina–Noyola, and J. L. Arauz–Lara, Phys. Rev. E **52** (1995) 4035.

[3] I. C. Kim and S. Torquato, J. Chem. Phys. **96** (1992) 1498.

[4] S. G. J. M. Kluijtmans, E. H. A. de Hoog, and A. P. Philipse, J. Chem. Phys. **108** (1998) 7469.

[5] A. Imhof and J. K. G. Dhont, Phys. Rev. **E52** (1995) 6344.

[6] P. N. Pusey, in *Liquids, Freezing and Glass Transition* , edited by J. P. Hansen, D. Levesque and J. Zinn–Justin (Elsevier, 1991).

[7] D. Stauffer and A. Aharony, *Introduction to Percolation Theory*, (Taylor and Francis, 1992).

[8] A. R. Mehrabi and M. Sahimi, Phys. Rev. Lett. **82**, 735 (1999).

[9] J. C. Crocker and D. G. Grier, J. Colloid Interface Sci. **179**, 298 (1996).

[10] M.D. Carbajal–Tinoco, G. Cruz de León, and J. L. Arauz–Lara, Phys. Rev. **E56** (1997) 6962.

[11] J. Santana–Solano and J. L. Arauz—Lara, Phys. Rev. E **65** (2002) 021406.

[12] D. A. McQuarrie, *Statistical Mechanics* (Harper and Row, New York, 1976).

SHORT-TIME TRANSPORT PROPERTIES IN COLLOIDS

O. Alarcón–Waess

Departamento de Física y Matemáticas, UDLA, Puebla,
Santa Catarina Martir, Cholula 72820 Puebla, México.

Abstract   We provide a method to perform transport properties, induced by an external field, in concentrated colloids at short–times. Hydrodynamic and direct interactions are considered. The idea is based in the appropriate choice of a slow variable, which multiplied by the external field is a contribution to the Hamiltonian of the colloid. We show that the transport coefficient, induced by the external field, is proportional to a generalized hydrodynamic function for very small wave vectors. To perform the generalized hydrodynamic function we assume that the time evolution of the slow variable is provided by the Smoluchowski equation.

Keywords: Hydrodynamic, colloid, transport properties

1.   INTRODUCTION

When one colloidal particle is in the presence of the action of an external field, the result is that the colloidal particle attains a certain velocity in the direction of the external field [1]. The most common example of an external field is the earth's gravitational ($\mathbf{E}_g$), in this case the phenomenon is called sedimentation. When the sedimentation flux of colloidal particles is allowed to proceed over an extended period of time, the so–called diffusion–sedimentation equilibrium is establish. Concentrations gradients then exists, such that the sedimentation flux of colloidal particles is counter balanced by gradient diffusion.

This equilibrium results in a transport property, which is the rate between the velocity and the external field. Another well known example of transport property is the electrophoretic mobility, which is related to the electric field.

The problem of the computation of transport properties, such as the sedimentation coefficient, of concentrated colloidal suspension from a molecular point of view is a very difficult task, because it deals with a

Developments in Mathematical and Experimental Physics,
Volume B: Statistical Physics and Beyond
Edited by Macias *et al.*, Kluwer Academic/Plenum Publishers, 2003

27

complex strongly interacting many particles system. The concentration dependence of transport properties at short–times of colloidal particles arises from both hydrodynamic and potential interactions between and individual particles as well as neighbors.

One important insight into the calculation of transport properties is due to Russel and Glendinning [2]. These authors point out that the hydrodynamic function $H(q)$ is related to the sedimentation coefficient in the $q \to 0$ limit. Further calculations have shown the importance of this relationship applied to colloids of spherical charged particles [3]. In this case the sedimentation coefficient also holds for long times if the collective memory kernel vanishes faster than q^2. This is strictly valid for monodisperse systems, provided that the direct forces and the hydrodynamic interactions are pairwise additive [4].

A new application of the idea of Russel and Glendinning was recently given by Alarcón–Waess and Diaz–Herrera [5] They computed the orientational hydrodynamic function $H(q)_{R,T}$, rotational and translational, for a dipolar spherical colloid. In this approach they defined the ordering coefficients, which are a measure of the orientational order, translational as well as rotational, in a dipolar colloid.

The aim of this paper is to generalize these ideas of Russel and Glendinning in order to compute short–time transport coefficients in colloids. In Section II we give the ideas of Russel and Glendinning, using the sedimentation phenomenon. In Section III we develop a general method to perform transport properties. In Section IV concluding remarks are provided.

2.   SEDIMENTATION

In order to analyze sedimentation, we consider a hydrodynamic description of the colloid, in which the slow variable is the normalized density fluctuations $\delta n(q,t)$ and our fast variable is the flux of colloidal particles $\mathbf{J}_n(q,t)$, where q is the magnitude of the wave vector, and t represents time.

The main quantity to describe the collective dynamics of colloids is the dynamic structure factor $F(q,t)$, which is the autocorrelation function of the normalized density fluctuations $\delta n(q,t)$, i.e.,

$$F(q,t) = < \delta n(q,t)^* \delta n(q,0) > . \qquad (1)$$

To compute the short–time behavior of the dynamic structure factor we assume that the temporal evolution of $\delta n(q,t)$ is provided by the Smoluchowski equation. The first cumulant of the dynamic structure factor at short–times is given by

$$\Gamma(q) = - \lim_{t \to 0} \left[\frac{\partial}{\partial t} \ln F(q,t) \right], \tag{2}$$

where $t \to 0$ means $\tau_B \ll t \ll \tau_I$, according to the Smoluchowski's temporal scale, and τ_B and τ_I are the Brownian and the structural relaxation time, respectively [4]. The importance of $\Gamma(q)$ is that it is related to the short–time collective diffusion coefficient ($D^{c,short}(q)$) [1, 4] in the following form

$$\Gamma(q) = q^2 D^{c,short}(q), \tag{3}$$

and the hydrodynamic function is given in terms of the short–time collective diffusion coefficient by the following relationship

$$H(q) = D^{c,short}(q) F(q) \tag{4}$$

where $F(q)$ is the static structure factor [1, 4]. On the other hand the hydrodynamic function can be written as

$$H(q) = \left\langle \frac{1}{N} \sum_{l,m}^{N} \hat{\mathbf{q}} \cdot \mathbf{D}_{lm}(\mathbf{R}^N) \cdot \hat{\mathbf{q}} \, e^{i \, \hat{\mathbf{q}} \cdot \mathbf{R}_l - \mathbf{R}_m} \right\rangle, \tag{5}$$

where $\mathbf{D}_{lm}(\mathbf{R}^N)$ are the many–body microscopic diffusion tensors [6]. The hydrodynamic function contains the configuration effect of the hydrodynamic interactions onto the short–time dynamics. Without hydrodynamic interactions $H(q)$ reduces to the free diffusion coefficient D^0.

Russel and Glendinning have pointed out that $H(q)$ can be interpreted as a sedimentation coefficient. In order to understand its interpretation we focus ourselves on the continuity equation, without earth's gravitational force,

$$\frac{\partial}{\partial t} \delta n(q,t) = i\mathbf{q} \cdot \mathbf{J}_n(q,t), \tag{6}$$

where $\mathbf{J}_n(q,t)$ can be written as

$$\mathbf{J}_n(q,t) = -M[\delta n(q,t)]\mathbf{q} \ln\{[\delta n(q,t)]\} \frac{d\Pi[\delta n(q,t)]}{d\delta n(q)} \delta n(q,t) \tag{7}$$

where $M[\delta n(q,t)]$ is the density dependent mobility, and $\Pi[\delta n(q,t)]$ is the osmotic pressure [1]. The mobility is defined as the proportionality constant between the velocity and the total force. Eq. (7) can be written in terms of the effective diffusion coefficient as

$$\mathbf{J}_n(q,t) = -iD[\delta n(q,t)]\mathbf{q}\delta n(q,t), \tag{8}$$

where $D[\delta n(q,t)]$ is the effective diffusion coefficient. For very weak inhomogeneities $D[\delta n(q,t) \sim D(\rho)$, which is precisely the gradient diffusion coefficient, where $\rho(= N/V)$ is the bulk density.

In presence of the gravitational force the total flux has a second contribution which gives rise to the phenomenon called sedimentation,

$$\mathbf{J}_s(q,t) = -M[\delta n(q,t)]\delta n(q,t)\mathbf{E}_g, \qquad (9)$$

where the total force is composed for the two contributions $\mathbf{q}\ln\{[\delta n(q,t)]\}$ $\frac{d\Pi[\delta n(q,t)]}{d\delta n(q)}$ and $\delta n(q,t)\mathbf{E}_g$. Therefore at $\mathbf{q} = 0$ the velocity will only be due to the earth's gravitational force, the sedimentation velocity. We consider a sedimentation–diffusion equilibrium where the sedimentation Peclet number is so small that the concentration and its gradients are small everywhere, so that a first order in concentration consideration suffices. In equilibrium the two forces will be equals so that

$$\mathbf{q}\delta n(q) = \frac{\rho\mathbf{E}_g}{\frac{d\Pi}{d\rho}} = \beta F(q \to 0)\mathbf{E}_g, \qquad (10)$$

where we have used that $F(q \to 0) = 1/(\beta d\Pi/d\rho)$. For the gradient diffusion coefficient,

$$D(\rho) = M(\rho)\frac{k_B T}{F(q \to 0)}, \qquad (11)$$

while from Eq. (4) we have

$$M(\rho) = \frac{H(q \to 0)}{D^0\beta}. \qquad (12)$$

Thus, using Eqs. (10), (11) and (12) we have that the hydrodynamic function is the sedimentation coefficient. This idea has been applied to hard spheres as well as charged spheres with interesting results [1, 4].

3.   TRANSPORT PROPERTIES

In this section we generalize the idea of Russel and Glendinning to provide a method to compute transport properties for short times. The aim is to compute the velocity induced by the presence of an external field $\mathbf{E}$, which we simply call "phenomenon velocity". We assume that a slow variable $Q(q,t)$ exists which when multiplied by $\mathbf{E}$ is the generalized external force. The time evolution of $Q(q,t)$ is provided by the Smoluchowski equation, which defines a precise temporal scale. The generalization of the idea of Russel and Glendinning consists in the choice of $Q(q,t)$. We know that the generalized velocity of colloidal material is proportional to the total generalized force, that is,

$$\mathbf{v}_g(q,t) = M[Q(q,t)]\mathbf{F}^{total}, \tag{13}$$

where $M[Q(q,t)]$ is the mobility and $\mathbf{F}^{total}$ is the total force, where the latter is composed by the external field and the gradients in the appropriate space (Fourier, angular, etc.) of $Q(q,t)$. From Eq. (13), we see that in the $q \to 0$ limit that the generalized velocity will only be induced by the external field. We define the phenomenon velocity at short times as

$$\mathbf{v}_p = M[Q(q=0)]\mathbf{E}. \tag{14}$$

It follows that to compute the transport property induced by the external force $Q(q,t)\mathbf{E}$, we first need to calculate the mobility. In order to perform $M[Q(q,t)]$ we will use an hydrodynamic description. We start by choosing the hydrodynamic variables, one of them is the slow variable $Q(q,t)$ and the other is the flux ($\mathbf{J}_Q(q,t)$) related to it . These variables must satisfy the continuity equation, that is

$$\frac{\partial}{\partial t}Q(q,t) + \nabla \cdot \mathbf{J}_Q(q,t) = 0, \tag{15}$$

where ∇ represents the gradient in the appropriate space (Fourier, angular, etc.).

The derivation of a model for the flux is based on the fact that in a dense colloid the relaxation of the flux is very fast and only the slow variable $Q(q,t)$ is important on the temporal scale of interest [7]. In our model we assume that the flux is driven by the corresponding gradient in the slow variable. For small gradients in $Q(q,t)$, the flux is a linear function of the gradient, which can thus formally be written as

$$\mathbf{J}_Q(q,t) = -\int dt' D(q,t-t')\nabla Q(q,t), \tag{16}$$

where the integral kernel $D(q,t)$ will be referred simply as the "phenomenon diffusion" coefficient, which is function of $Q(q,t)$. In the presence of the external field $\mathbf{E}$, then the total flux in Eq. (15) has an additional component, induced by $\mathbf{E}$, which is written as

$$\mathbf{J}_E(q,t) = M[Q(q,t)]Q(q,t)\mathbf{E}. \tag{17}$$

In order to compute the transport property, we assume very weak inhomogeneities, i.e., the flux $\mathbf{J}_Q(q,t)$ is given by the Fick's law, which holds for short times, so Eq. (16) has the simplest form

$$\mathbf{J}_Q(q,t) = -D_\nabla \nabla Q(q,t), \tag{18}$$

where D_∇ is the generalized gradient diffusion which describes the transport of colloidal particles in a slow variable $Q(q,t)$ with a constant gradient. For very small wave vectors the short–time collective phenomenon diffusion coefficient is equal to the corresponding generalized gradient diffusion coefficient,

$$D_\nabla = \lim_{q \to 0} D^{c,short}[Q(q)]. \qquad (19)$$

We consider a "gedanken experiment" in which the generalized diffusion-phenomenon stationary state is established, that is the flux driven by the external field is counter balanced by the flux of the generalized gradient diffusion. From Eqs. (17), (18) and (19), the mobility $M[Q(q = 0)]$ must be proportional to phenomenon hydrodynamic function $H_p(q = 0)$. Where the latter is the hydrodynamic function given in Eq. (4), with the difference that the collective diffusion coefficient is computed using $Q(q,t)$ in the dynamic structure factor. As a consequence of this, the phenomenon hydrodynamic function is proportional to the phenomenon coefficient at $q = 0$. Therefore, we obtain for the short time regime

$$\frac{v_p}{v_{p0}} = \frac{H_p(q = 0)}{D^0}, \qquad (20)$$

where v_{p0} is the phenomenon velocity for an isolated colloidal particle in the same solvent.

An example of this method was recently applied to a dipolar colloid [5]. In this case the slow variable was the orientation density. Applying this method the authors obtained two transport properties, called by them the translational and rotational ordering coefficients. The external force in this case was the angular gradient of the orientation density. The importance of these coefficients is that they are related with the angular gradient of the orientation density if a dipolar colloid lose some positional and/or orientational symmetries induced by.

4.   CONCLUDING REMARKS

We have generalized the ideas of Russel and Glendinning to perform transport properties for short times, considering hydrodynamic and direct interactions. The short time is defined by the time scale of the Smoluchowski equation. Our method provides a simple way to perform a transport coefficient which is induced by an external field. We have shown that this transport property is related to a generalized hydrodynamic function, which is computed using an appropriate slow variable.

The method provided consist in choosing an appropriate slow variable. The first step is to write the Hamiltonian of the colloid, in which the contribution to the Hamiltonian due to the external field indicates which slow variable is appropriate. The next step is to compute the dynamic structure factor using this slow variable, and the final step is to compute its first cumulant which is related to the generalized hydrodynamic function, using the slow variable previously defined. It is important to realize that the time evolution of the slow variable is provided by the Smoluchowski equation.

An open question concerns long times transport properties. In some cases the collective diffusion coefficient is time independent for very small wave vectors. In this case our result is also valid for long times. One condition is that the collective memory kernel vanishes faster than q^2.

The main value of this method is that it is very simple, and the new transport coefficient can predict the response of a concentrated colloid in the presence of an external field.

Acknowledgments

This work was supported by funds from CONACYT, Mexico, Grant No. 28239 E.

References

[1] J. K. G. Dhont, An Introduction to Dynamics of Colloids (Elsevier, New York, 1996)

[2] W. B. Russel and A. B. Glendinning, J. Chem. Phys. **74** (1981) 948.

[3] R. Klein and G. Naegele, Il Nuovo Cimento **16** (1994) 963.

[4] G. Naegele, Phys. Rep. **272** (1996) 215.

[5] O. Alarcón–Waess and E. Diaz–Herrera, Phys. Rev. **E65** (2002) 031402.

[6] R. Schmitz and B. U. Felderhof, Physica **A116** (1982) 163.

[7] B. Bagchi and A. Chandra, in Advances in Chemical Physics, edited by I. Prigogine and S. A. Rice (Whiley, New York, 1991), Vol. LXXX.

ROTATIONAL DIFFUSION IN A FERROCOLLOID

M. Hernández–Contreras

Departamento de Física

Centro de Investigación y Estudios Avanzados del Instituto Politécnico Nacional

A.P. 14–740, México D.F., México.

marther@fis.cinvestav.mx

Abstract   The longtime rotational self–diffusion coefficient in a ferrofluid was derived from a Langevin equation approach. This property depends on the micro-structure of the bulk solution, and was determined through a linearized hypernetted chain approximation. It is shown there is a strong reduction of the rotational diffusion due to interparticle's interactions.

Keywords: Diffusion, colloid, ferrofluid

1.   INTRODUCTION

Despite two decades of intensive research on ferrocolloids, there remains many open questions on the dynamic properties in these magnetic fluids [1]. Most of the studies on these materials are focused on the phase behavior and their microstructure [2]. A ferrofluid is a colloidal dispersion of magnetic particles in a carrier fluid. They bear an intrinsic birefringence anisotropy which allows a very accurate optical detection of their rotations [3]. Experimentally, the techniques of x–ray, neutron, and depolarized light–scattering measure the long–time tracer rotational diffusion coefficient D_R of their rotations at thermal equilibrium [4]. In this paper we determine in Section 1 the rotational diffusion coefficient in a concentrated ferrofluid from a generalized Langevin equation approach. The expression for this dynamical property depends on the equilibrium microstructure of the dispersion given by the total correlation function h of the bulk solution. For such a suspension we derive in Section 2 an accurate h from a linearized hypernetted chain (LHNC) liquid theory. In Section 3 it is given an analysis of the rotational diffusion

Developments in Mathematical and Experimental Physics,
Volume B: Statistical Physics and Beyond
Edited by Macias *et al.*, Kluwer Academic/Plenum Publishers, 2003

35

coefficient as determined from the microstructural properties. Section 4 contains a brief conclusion.

2.   LANGEVIN EQUATION APPROACH

The translational and rotational Brownian motion of a tracer dipolar particle that diffuses in a suspension of other anisotropic particles with which is interacting, is given by [5]

$$\overleftrightarrow{\mathrm{M}} \cdot \frac{d\vec{\mathbf{V}}(t)}{dt} = -\overleftrightarrow{\zeta^0} \cdot \vec{\mathbf{V}}(t) + \vec{\mathbf{f^0}}(t)$$

$$- \int_0^t dt' \Delta\overleftrightarrow{\zeta}(t - t') \cdot \vec{\mathbf{V}}(t') + \vec{\mathbf{F}}(t). \tag{1}$$

Here $\vec{\mathbf{V}}(t) = (\mathbf{V}(t), \mathbf{W}(t))$ embodies the linear $\mathbf{V}(t)$ and angular $\mathbf{W}(t)$ velocities of the tracer, whose components are defined with respect to a coordinate system with origin fixed to the laboratory frame, but whose orientation changes with time following the orientation of the main axis of symmetry of the tracer. We consider the dipolar anisotropy to be oriented in the direction of the z axis of the reference frame. $\overleftrightarrow{\mathrm{M}}_{ij} = M\delta_{ij}$ $(i, j = 1, 2, 3)$, $\overleftrightarrow{\mathrm{M}}_{ij} = \delta_{ij}I_{i-3}$ $(i, j = 4, 5, 6)$, with M, I_1, I_2, I_3 being the mass and principal moments of inertia of the tracer; and $\overleftrightarrow{\zeta^0}_{ij}$ $(i, j = 1, 2, ..., 6)$, that turns out to be diagonal, are the friction coefficients coupling the random force $\mathbf{f}(t)$ and torque $\mathbf{T}(t)$, grouped in $\vec{\mathbf{f^0}}(t)$, with the generalized velocity $\vec{\mathbf{V}}(t)$. This friction tensor has the nonzero components $\zeta_{11}^0 = \zeta_{22}^0 = \zeta_\perp^0$, $\zeta_{33}^0 = \zeta_\parallel^0$, $\zeta_{44}^0 = \zeta_{55}^0 = \zeta_R^0$, and $\zeta_{66}^0 = 0$. In Eq. (1), $\vec{\mathbf{F}}(t)$ is a fluctuating generalized force deriving from the spontaneous departures from zero of the net direct force exerted by the other particles on the tracer. It groups a random force and torque on the tracer with zero mean value, and time dependent correlation function given by $\langle \vec{\mathbf{F}}(t)\vec{\mathbf{F}}^\dagger(0) \rangle = k_\mathrm{B}T\Delta\overleftrightarrow{\zeta}(t)$, where the time–dependent friction function is

$$\Delta\zeta_\gamma(t) = \beta \int d^3\mathbf{r} d^3\mathbf{r}' d\Omega d\Omega' \, [\nabla_\gamma \psi(\mathbf{r}, \Omega)]$$

$$\times C(\mathbf{r}, \mathbf{r}', \Omega, \Omega'; t) \, [\nabla'_\gamma \psi(\mathbf{r}', \Omega')], \tag{2}$$

with $\beta = 1/k_\mathrm{B}T$, $\Omega = (\phi, \theta)$, and θ, ϕ are the polar angles.

The time–dependent memory function $\Delta\zeta_\gamma(t)$ contains the dissipative friction effects derived from the direct interactions of the tracer with the particles around it. This memory function defines the relaxation time $\tau_\mathrm{I} \gg \tau_\mathrm{B}$ (τ_B is the relaxation time of the momenta of the particles) for the particles to diffuse a mean distance among them. Thus, in the diffusive regime $t \gg \tau_B$, long times mean $t \gg \tau_I$. The generalized gradient operator is given by $\nabla_\gamma = (\nabla, \mathbf{r} \times \nabla + \nabla_\Omega)_\gamma$, where ∇_Ω is the angular gradient operator [6]. For axial symmetric potentials $\nabla_\Omega = \mathbf{n} \times d/d\mathbf{n}$ with $\mathbf{n}$ the unit cartesian vector in the direction of the axis of symmetry of the particle. In Eq. (2), $C(\mathbf{r}, \mathbf{r}', \Omega, \Omega'; t) \equiv \langle \delta n(\mathbf{r}, \Omega; t) \delta n(\mathbf{r}', \Omega'; 0) \rangle$ is the van Hove function of the fluctuations in the concentration of particles with respect to the equilibrium value $n(\mathbf{r}, \Omega; t)$, and determines the relaxation modes of the tracer and of the cage of particles surrounding it. Its proper definition takes into account that it is referred to the tracer position and orientation, and has the initial condition value $C(t = 0) = \langle \delta n(0) \delta n(0) \rangle \equiv \sigma$, i.e., the static correlation function, whose inverse σ^{-1} is defined by

$$\int d\mathbf{r}'' \, d\Omega'' \, \sigma(\mathbf{r}, \mathbf{r}'', \Omega, \Omega'') \, \sigma^{-1}(\mathbf{r}'', \mathbf{r}', \Omega'', \Omega') =$$
$$\delta(\mathbf{r} - \mathbf{r}')\delta(\Omega - \Omega').$$

Eq. (2) for $\Delta\zeta_\gamma(t)$ is an exact result of the GLE approach for the description of the effect on the translational and rotational diffusion of the tracer particle due to the direct interactions with the cloud of nonspherical particles that surround it. We approximate the properties σ and $C(t)$ by their values far from the tracer particle, i.e., we introduce the homogeneity approximation. Then, we also assume that $C(t)$ is given by the solution of the general linearized Fick's diffusion equation

$$
\begin{aligned}
\frac{\partial C(\mathbf{r} - \mathbf{r}', \Omega, \Omega'; t)}{\partial t} &= -\rho[D^0 \nabla^2 + D_R^0 \nabla_\Omega^2] \\
&\quad \times \int dt' d^3\mathbf{r}'' d\Omega'' \sigma^{-1}(\mathbf{r} - \mathbf{r}'', \Omega, \Omega'') \\
&\quad \times C(\mathbf{r}'' - \mathbf{r}', \Omega'', \Omega'; t - t'),
\end{aligned}
\tag{3}
$$

where ρ is the bulk concentration of particles, whereas D^0 and D_R^0 are phenomenological parameters that should be provided by experiment or an external theory. We are interested in determining the departure of the long–time diffusion coefficient $D_R^0 = k_B T / (\zeta_R^0 + \Delta\zeta_R)$ due to interparticle interactions, from the its free diffusion value $\zeta_R^0 = k_B T / D_{R}^0$, and derived in the diffusive regime $t \gg \tau_I$, from

$$\Delta\overset{\leftrightarrow}{\zeta} = \int_0^\infty dt \, \Delta\overset{\leftrightarrow}{\zeta}(t). \tag{4}$$

3. MICROESTRUCTURE OF THE FERROFLUID

We consider the ferrocolloid to be formed by a monodisperse suspension of same size spherical particles of diameter d with an equal and constant dipole moment of strength μ located at their center. Thus, the pair interaction between two particles a given distance r apart is

$$\psi(\mathbf{r}, \Omega_1, \Omega_2) = u_{HS}(r) + u^{112}(r) \tag{5}$$

where $u_{HS}(r)$ is the hard sphere interaction contribution, and $u^{112}(r) = \mu^2 \left[3(\hat{\mathbf{r}} \cdot \mathbf{n}_1)(\hat{\mathbf{r}} \cdot \mathbf{n}_2) - (\mathbf{n}_1 \cdot \mathbf{n}_2) \right] / r^3$, where $r = |\mathbf{r}|$, $\hat{\mathbf{r}} = \mathbf{r}/r$ is the unitary vector with direction Ω_r, and $\hat{\mathbf{n}}_i = \vec{\mu}_i / \mu$. For the axialsymmetric pair potential given above we use the rotational invariant expansion [7]. Thus, for any function $f(\mathbf{r}, \Omega, \Omega')$, for example $C(t)$, this expansion is given by

$$
f(\mathbf{r}, \Omega, \Omega') = (4\pi)^{3/2} \sum_{mnl} \frac{f^{mnl}(r)}{\sqrt{2l+1}} \sum_{\mu'\nu'\lambda'} \begin{pmatrix} m & n & l \\ \mu' & \nu' & \lambda' \end{pmatrix}
$$
$$
\times Y_{m\mu'}(\Omega) Y_{n\nu'}(\Omega') Y_{l\lambda'}(\Omega_r), \tag{6}
$$

where we have used the usual notation for the $3j$ symbols, the $Y_{l\lambda}$ are the spherical harmonics, and Ω_r is the angle of the orientation of the position vector $\mathbf{r}$.

The symmetries of the dipolar potential imposes to its invariant representation of Eq. (6) the minimal basis set : (mnl)=(000), (110), (112). It is known that this basis leads to an accurate description of the microstructural total correlation function h of the suspension when determined through the LHNC theory [8]. It compares well to Monte Carlo calculations for the same model interaction and thermodynamic states [8]. In the LHNC the corresponding $h^{000}(r), h^{110}(r)$ and $h^{112}(r)$ are obtained using the exact $h^{000}(r)$ from a hard sphere system according to the Verlet–Weiss method, and the Ornstein–Zernike's equations given in Fourier space by

$$\gamma^{000}(k) = \rho^* c^{000}(k) h^{000}(k) \tag{7}$$

$$\gamma^{112}(k) = \frac{\rho^*}{3} \left[c^{112}(k) h^{112}(k) + c^{112}(k) h^{110}(k) + c^{110}(k) h^{112}(k) \right] \tag{8}$$

$$\gamma^{110}(k) = \frac{\rho^*}{3} \left[2c^{112}(k) h^{112}(k) + c^{110}(k) h^{110}(k) \right] \tag{9}$$

where $\gamma^{mnl}(k) = h^{mnl}(k) - c^{mnl}(k)$, c^{mnl} is the direct correlation function, and $c^{mnl}(k) = 4\pi i^l \int_0^\infty dr\, r^2\, j_l(kr)\, c^{mnl}(r)$, with $j_l(kr)$ the spherical Bessel function, and $\rho^* = \rho d^3$. The solutions of Eqs. (7–9) are obtained from the LHNC approximation

$$c^{000}(r) = h^{000}(r) - \ln g^{000}(r) - \beta u_{HS}(r) \tag{10}$$

$$c^{mnl}(r) = g^{000}(r)\left[\gamma^{mnl}(r) - \beta u^{mnl}(r)\right] - \gamma^{mnl}(r) \tag{11}$$

with the pair correlation function $g^{mnl}(r) = h^{mnl}(r) + 1$ defined in real space.

In Fig. 1(a) and 1(b) are shown the results for the components $h^{112}(r) = \sqrt{30}h_D(r)$ and $h^{110}(r) = -\sqrt{3}h_\Delta(r)$, respectively, as a function of interparticle separation r and fixed volume fraction $\phi = .41$ and reduced dipole strength $\mu^2 = \beta\mu^{*2}/d^3 = 0.172$.

From the microstructural properties given by $h^{000}(r)$, $h^{110}(r)$, $h^{112}(r)$ we defined the conventional hard sphere, longitudinal and transverse dipole, structure factors, respectively

$$\begin{aligned}
S_{,0}^{00}(x) &= 1 + \frac{6\phi}{\pi}h^{000}(x), \\
S_{,0}^{11}(x) &= 1 - \frac{6\phi}{\pi}\left(\frac{1}{\sqrt{3}}h^{110}(x) - \frac{2}{\sqrt{30}}h^{112}(x)\right), \\
S_{,\pm1}^{11}(x) &= 1 - \frac{6\phi}{\pi}\left(\frac{1}{\sqrt{3}}h^{110}(x) + \frac{1}{\sqrt{30}}h^{112}(x)\right),
\end{aligned} \tag{12}$$

where $x = kd$, and use was made of the definition

$$f_{,\chi}^{mn}(k) = \sum_{l=|m-n|}^{m+n} \begin{pmatrix} m & n & l \\ \chi & -\chi & 0 \end{pmatrix} f^{mnl}(k). \tag{13}$$

4.   ROTATIONAL DIFFUSION COEFFICIENT

In this section we determine the friction coefficient. Use of the Fourier and Laplace transforms in Eq. (3) of $C(t)$, in combination with the invariant expansion, Eq. (6), the minimal basis set for (mnl), it is straightforward to obtain a closed expression for the $(11,\chi)$ components of the frequency–dependent collective diffusion propagator which reads

$$C_{,\chi}^{11}(k,w) = \frac{C_{,\chi}^{11}(k,t=0)}{-iw + \rho 4\pi(-)^\chi(D^0 k^2 + 2D_R^0)[\sigma^{-1}(k)]_{,\chi}^{11}}. \tag{14}$$

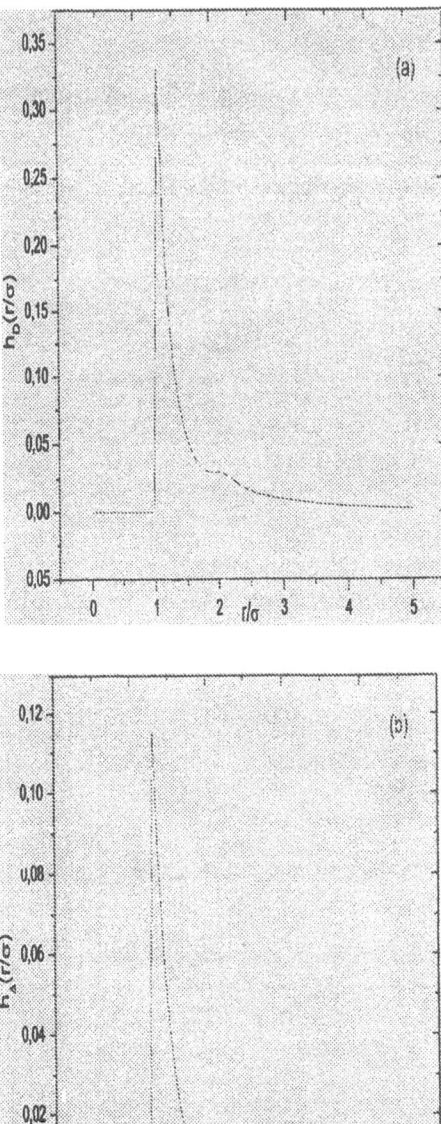

Figure 1. Total correlation function $h_D(kd) = h^{112}(kd)/\sqrt{30}$: (a), $h_\Delta(kd) = -h^{110}(kd)/\sqrt{3}$: (b), for a volume fraction $\phi = 0.41$ and reduced dipolar strength $\mu^{*2} = 0.172$.

Then, this form of $C(w)$ is used in Eqs. (2) and (4) to get the corresponding expressions of $\Delta\zeta_\gamma$. Thus, the longtime translational friction ($w = 0$) reads

$$\Delta\zeta_\gamma(w = 0) = \Delta\zeta^{HS} + \Delta\zeta^p_\gamma, \qquad (15)$$

for $\gamma = \perp, \parallel$, where

$$\Delta\zeta^{HS} = \frac{\zeta^0}{288\pi^2\phi} \int_0^\infty dx \; x^2 \left(S^{00}_{,0}(x) - 1\right)^2 \qquad (16)$$

is the contribution from the hard sphere core [9], and

$$\Delta\zeta^p_\perp = \frac{576}{5} \zeta^0 \phi\mu^{*4} \int_0^\infty dx \; \frac{j_2(x)^2}{x^2 + 3/2} \\ \times \left[2S^{11}_{,1}(x)^2 + 3S^{11}_{,0}(x)^2\right] \qquad (17)$$

whereas the parallel translational friction reads

$$\Delta\zeta^p_\parallel = \frac{4}{3}\Delta\zeta^p_\perp. \qquad (18)$$

The above equations for $\Delta\zeta_\gamma$ give the friction contributions from the dipole–dipole interactions. The parameter ζ^0 is the translational friction coefficient for a solid sphere in a fluid of viscosity η, and is given by the Stokes formula $\zeta^0 = 6\pi\eta d$. In the same fashion, the longtime rotational friction gives

$$\Delta\zeta_R = \frac{48}{135} \zeta^0_R \phi\mu^{*4} \int_0^\infty dx \; \frac{j_1(x)^2}{x^2 + 3/2} \\ \times \left[133S^{11}_{,1}(x)^2 + 164S^{11}_{,0}(x)^2\right]. \qquad (19)$$

where ζ^0_R is the rotational friction coefficient for a spherical particle, and is given by $\zeta^0_R = 8\pi\eta d^3$.

In Fig. 2 it is depicted the normalized rotational self–diffusion coefficient D_R/D^0_R as a function of the reduced dipole moment μ^{*2} and fixed volume fraction $\phi = .41$.

As it is expected, the rotational diffusion of the particles is diminished when the dipole moment increases. Another interesting case is obtained when ϕ is varied while μ^{*2} is kept to a constant value, it is also found a strong reduction of D_R/D^0_R (not depicted). It would be interesting to compare the prediction of the simple theoretical scheme based on Eqs. (17-19) presented here for the rotational diffusion in ferrofluids both, with model computer simulations, or with the diffusion coefficient obtained experimentally from a real ferrofluid. Such a comparison have being made for the long–time center of mass translational

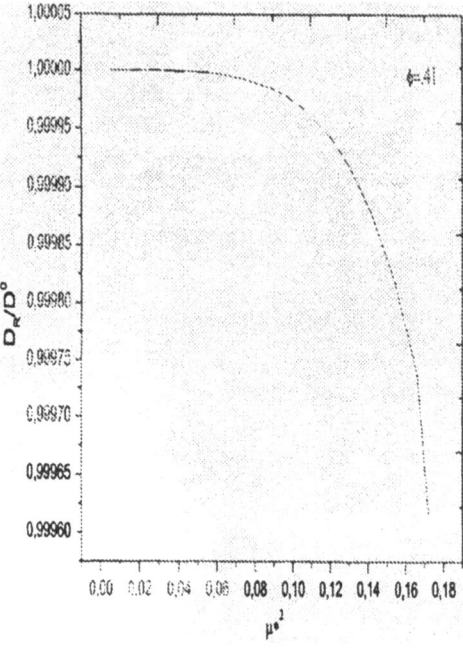

Figure 2. D_R/D_R^0 as a function of μ^{*2}, and fixed volume fraction $\phi = 0.41$

self–diffusion coefficient in a highly charged concentrated suspension of spherical polyions and compared with a theory for the same diffusion property [9]. Good agreement was found in that case [10]. We expect good qualitative agreement could be reached for the rotational diffusion coefficient given here, Eq. (19), when compared to its experimental counterpart [11].

5.   CONCLUSION

We determined the rotational self–diffusion coefficient of ferrofluid particles using a Langevin equation approach. The diffusion coefficient depends on the static structural properties of the bulk solution as given by the dipolar interactions among particles. Accurate structural properties were obtained from a linearized hypernetted chain approximation at equilibrium liquid states and the diffusion coefficient was calculated correspondingly. It is found that dipolar interactions suppress the rotational self–diffusion.

Acknowledgments

This work was supported by a Conacyt grants 32169–E and 36557–E, México.

References

[1] G.A. Flores *et al*, in: *Proceedings of the Fifth International Conference on ER Fluids, MR Suspensions, and Associated Technology*, ed. W. Bullough, Shieffield (World Scientific, Singapore, 1996).

[2] S.H.L. Klapp and G.N. Patey, *J. Chem. Phys.* **115** (2001) 4718.

[3] R. Piazza, V. DeGiorgio, M. Corti, and J. Stavans, *Phys. Rev. B* **42** (1990) 4885.

[4] B.J. Berne and R. Pecora, *Dynamic Light Scattering*, ed. B.J. Berne and R. Pecora, Wiley, New York, 1975.

[5] M. Hernández–Contreras and M. Medina–Noyola, *Phys. Rev.* **E54** (1996) 6573.

[6] *Theory of Molecular Fluids Vol.1*, ed. C.G. Gray and K.E. Gubbins, Clarendon Press, Oxford 1984.

[7] L. Blum, *J. Chem. Phys.* **57** (1972) 1862.

[8] P.H. Fries and G.N. Patey, *J.Chem.Phys.* **82** (1985) 429.

[9] M. Medina–Noyola, *Phys. Rev. Lett.* **60** (1988) 2705.

[10] G. Nagele, *Phys. Rep.* **272** (1996) 1.

[11] Work in progress.

DYNAMICS OF COLLOIDAL PARTICLES

J. M. Méndez–Alcaraz.
Depto. de Física, CINVESTAV–IPN, Av. IPN 2508, Col. San Pedro Zacatenco, 07360 México D. F., México.

O. Alarcón–Waess.
Depto. de Física y Matemáticas, UDLA, Puebla, Sta. Catarina Mártir, Cholula, 72820 Puebla, México.

Abstract   We present a general method for constructing effective one particle dynamics in colloidal systems, which results from a contraction of the description of colloidal mixtures based on the many particles Langevin and Smoluchowski equations. In order to illustrate its applicability, we calculate the long–time self-diffusion coefficient of charged particles, as well as its electrophoretic mobility. The accuracy of our approaches is pointed out by comparison with available experimental data.

Keywords: Electrophoresis, diffusion, colloids

1.   INTRODUCTION

The experimental access to complex fluids is always limited. For example, a clean salty aqueous suspension of polystyrene spheres contains obviously polystyrene spheres and water molecules, but also ions dissociated from the polystyrene molecules on the surface of the spheres, called counterions, and at least two different species of salt ions. The size of the first ones ranges from about some tens of nanometers, up to several microns. The other particles are of subnanometer dimensions. Observing the suspension by means of light scattering, only the polystyrene spheres are detected by the experiment, because they are the only particles able to affect the photons of micrometers wave lengths. The explanation of the experimental results requires therefore the theoretical formulation of effective models able to capture the effects induced by the unobserved particles on the physics of the observed ones. In our example, it leads to effective screened Coulombic pair interaction potentials between polystyrene particles, containing parameters related to the solvent

Developments in Mathematical and Experimental Physics,
Volume B: Statistical Physics and Beyond
Edited by Macias *et al.*, Kluwer Academic/Plenum Publishers, 2003   45

molecules and small ions. As explained in another paper in this book, it results from a contraction of the description of colloidal mixtures based on the integral equations theory of simple liquids [1]. The same integral equations theory accounts for the structural properties of the observed suspension by taking the effective interaction potentials as input [2].

The explanation of the dynamic properties of the observed particles is more complicated. They are immersed in an invisible medium composed by, for example, the unobserved solvent molecules and small ions, with which they exchange energy and momentum. Their dynamics is therefore non–conservative, which is a physical feature non-contained in the effective pair interaction potentials. In this paper we present a general method for constructing effective one particle dynamics in colloidal systems, which results from a further contraction of the description of colloidal mixtures based on the many particles Langevin and Smoluchowski equations. In order to illustrate its applicability, we calculate the long–time self–diffusion coefficient of charged particles, as well as its electrophoretic mobility. The accuracy of our approaches is tested by their comparison with available experimental data.

2.   CONTRACTION OF THE DESCRIPTION AND EFFECTIVE DYNAMICS

Let us think further about a salty aqueous suspension of polystyrene spheres. It means, a liquid mixture of five components or species; the polystyrene spheres or macroions (species m), the ions dissociated from the polystyrene molecules in the spheres or counterions (species c), the positive salt ions (species $+$), the negative salt ions (species $-$) and the water molecules (species w). Globally, the suspension is electroneutral. Locally, however, the system is charged and the particles interact between them with Coulombic potentials. The species w looks like electric dipoles (molecular model). In a first contraction of the description, the water molecules can be included only through the dielectric constant ϵ of the solvent, leading to the primitive model (PM) for the pair interaction potential $u_{\alpha\beta}(r)$ between two particles of species α and β separated by the distance r:

$$
\begin{aligned}
u_{\alpha\beta}(r) &= +\infty \quad \text{for} \quad r < a_\alpha + a_\beta \\
&= \frac{Q_\alpha Q_\beta}{\epsilon}\frac{1}{r} \quad \text{for} \quad r \geq a_\alpha + a_\beta,
\end{aligned}
\tag{1}
$$

with $\alpha, \beta = m, c, +, -$. Here, the hard core prevents the interpenetration and Q_α (a_α) is the charge (radius) of a particle of species α. A second contraction of the description can be carried out from the PM, so that the small ions no longer appear as separated species. This leads to the one component model (OCM) or effective screened Coulombic interaction between the macroions [1, 3, 4]:

$$u_{mm}(r) = +\infty \quad \text{for} \quad r < 2a_m$$
$$= \frac{Q_m^2}{\epsilon}\left(\frac{e^{\kappa a_m}}{1+\kappa a_m}\right)^2 \frac{e^{-\kappa r}}{r} \quad \text{for} \quad r \geq 2a_m, \qquad (2)$$

with

$$\kappa^2 = \frac{4\pi}{k_B T \epsilon} \sum_{\alpha=c,+,-} n_\alpha Q_\alpha^2 \qquad (3)$$

being the square inverse Debye–Hückel screening length and $n_\alpha = N_\alpha/V$ the number density of particles of species α in the volume V.

The structure of the suspension can be calculated either in the PM by solving the Ornstein–Zernike (OZ) equation [5]

$$h_{\alpha\beta}(r) = c_{\alpha\beta}(r) + \sum_{\gamma=m,c,+,-} n_\gamma \int_V c_{\alpha\gamma}(r')h_{\gamma\beta}(|\mathbf{r}-\mathbf{r}'|)d\mathbf{r}' \qquad (4)$$

or in the OCM by solving the more contracted OZ equation

$$h_{mm}(r) = c_{mm}(r) + n_m \int_V c_{mm}(r')h_{mm}(|\mathbf{r}-\mathbf{r}'|)d\mathbf{r}', \qquad (5)$$

together with an appropriate closure relation [5], by using (1) or (2) as input for the potential, respectively. Here, $c_{\alpha\beta}(r)$ is the direct correlation function, $h_{\alpha\beta}(r) = g_{\alpha\beta}(r) - 1$ the total correlation function and $g_{\alpha\beta}(r)$ the partial radial distribution function. Once $g_{\alpha\beta}(r)$ is known, all structural functions and thermodynamical properties of the system can be evaluated, each of them in their respective level of description [2, 5]. The macoion–macroion structure functions obtained by following both routes are the same, equations (1) and (4) or equations (2) and (5), since that is a necessary condition in the construction of the effective interaction potentials [1, 6].

In the PM, the motion of particle i of species α is described by the many particles Langevin equation

$$m_\alpha \frac{d\mathbf{v}_i^\alpha}{dt} = -\sum_\beta \sum_j \zeta_{ij}^{\alpha\beta} \cdot \mathbf{v}_j^\beta - \nabla_i U + \mathbf{f}_i^\alpha, \qquad (6)$$

which corresponds to the second Newton's law for a particle of mass m_α moving with velocity $\mathbf{v}_i^\alpha$ at time t. The force is the superposition of the friction $-\sum_\beta \sum_j \zeta_{ij}^{\alpha\beta} \cdot \mathbf{v}_j^\beta$ due to the solvent, the interaction $-\nabla_i U$ with the other particles and the Gaussian random force $\mathbf{f}_i^\alpha(t)$ related to $\zeta_{ij}^{\alpha\beta}$

through a fluctuation–dissipation theorem (FDT). The Greek indices refer to the species (m, c, $+$ or $-$) and the Latin ones to the particles (running from 1 to the total number of particles of the corresponding species N_m, N_c, N_+ or N_-); particle i is of species α and particle j of species β. The quantity U stands for the total interaction potential energy, assumed to be pair–wise additive.

An alternative description of the collective motion of the $N = \sum_\alpha N_\alpha$ particles in the system is the many particles Smoluchowski equation

$$\frac{\partial P^N}{\partial t} = \sum_{\alpha,\beta} \sum_{i,j} \nabla_i \cdot \mathbf{D}_{ij}^{\alpha\beta} \cdot \left[\nabla_j P^N + \beta \left(\nabla_j U \right) P^N \right], \qquad (7)$$

which corresponds to the Liouville equation (after integrating over all variables of the solvent molecules and the momentum of the polystyrene spheres and small ions [7]) for the normalized distribution function $P^N = P_{\alpha\beta\cdots\gamma}^N(\mathbf{r}_1, \mathbf{r}_2, \cdots, \mathbf{r}_N; t)$ of finding particles 1 of species α in $\mathbf{r}_1$, $\cdots$, and particle N of species γ in $\mathbf{r}_N$ at time t. The diffusion tensor $\mathbf{D}$ and the friction tensor ζ are connected by the Einstein relation $\mathbf{D} = k_B T \zeta^{-1}$. Both tensors contain the effects induced by the invisible solvent onto the dynamics of the observed particles, known as hydrodynamic interactions (HI). We take here the most simple approximation for them; $\zeta_{ij}^{\alpha\beta} = \zeta_\alpha \delta_{ij}^{\alpha\beta} \mathbf{1}$ and $\mathbf{D}_{ij}^{\alpha\beta} = D_\alpha \delta_{ij}^{\alpha\beta} \mathbf{1}$, where ζ_α is given by the Stokes expression $6\pi\eta a_\alpha$, η being the shear viscosity of the solvent.

In order to derive a closed equation for $\mathbf{v}_i^\alpha$ from equations (6) and (7), we make use of reduced distribution functions $\rho_{\alpha\beta\cdots\gamma}^{(n)}(\mathbf{r}_1, \mathbf{r}_2, \cdots, \mathbf{r}_n; t) = N_\alpha(N_\beta - \delta_{\beta\alpha}) \cdots (N_\gamma - \delta_{\gamma\alpha} - \delta_{\gamma\beta} - \cdots) \int P^N d\mathbf{r}_{n+1} \cdots d\mathbf{r}_N$ of finding any particle of species α in $\mathbf{r}_1$, $\cdots$, and any particle of species γ in $\mathbf{r}_n$ at time t, regardless of the rest of the particles. Carrying out this integration, equation (6) takes the continuous form

$$m_\alpha \frac{d\mathbf{v}_1^\alpha}{dt} = -\zeta_\alpha \mathbf{v}_1^\alpha - \frac{1}{N_\alpha} \sum_\beta \int_V \int_V [\nabla_1^\alpha u_{\alpha\beta}(r_{12})] \, \rho_{\alpha\beta}^{(2)}(\mathbf{r}_1, \mathbf{r}_2; t) d\mathbf{r}_1 d\mathbf{r}_2 + \mathbf{f}_1^\alpha.$$

$$(8)$$

In addition, equation (7) gives rise to a hierarchy of evolution equations involving all reduced distribution function. The member of the hierarchy corresponding to $\rho^{(2)}_{\alpha\beta}(\mathbf{r}_1, \mathbf{r}_2; t)$ becomes

$$
\begin{aligned}
\frac{\partial \rho^{(2)}_{\alpha\beta}}{\partial t} =\ & D_\alpha \nabla_1^2 \rho^{(2)}_{\alpha\beta} + D_\beta \nabla_2^2 \rho^{(2)}_{\alpha\beta} \\
& + \beta D_\alpha \nabla_1 \cdot \left[(\nabla_1 u_{\alpha\beta}(r_{12})) \, \rho^{(2)}_{\alpha\beta} \right] + \beta D_\beta \nabla_2 \cdot \left[(\nabla_2 u_{\alpha\beta}(r_{21})) \, \rho^{(2)}_{\alpha\beta} \right] \\
& + \beta D_\alpha \nabla_1 \cdot \sum_\gamma \int_V (\nabla_1 u_{\alpha\gamma}(r_{13})) \, \rho^{(3)}_{\alpha\beta\gamma} d\mathbf{r}_3 \\
& + \beta D_\beta \nabla_2 \cdot \sum_\gamma \int_V (\nabla_2 u_{\beta\gamma}(r_{23})) \, \rho^{(3)}_{\alpha\beta\gamma} d\mathbf{r}_3.
\end{aligned}
\tag{9}
$$

We now pass from the partial to the total time derivative by adding the streaming term $\mathbf{v}_1^\alpha \cdot \nabla_1 \rho^{(2)}_{\alpha\beta}$;

$$
\frac{d\rho^{(2)}_{\alpha\beta}}{dt} = \mathbf{v}_1^\alpha \cdot \nabla_1 \rho^{(2)}_{\alpha\beta} + \frac{\partial \rho^{(2)}_{\alpha\beta}}{\partial t}.
\tag{10}
$$

This completes the scheme, which can be recapitulated as following: First, an approximation for $\rho^{(3)}_{\alpha\beta\gamma}$ in terms of $\rho^{(2)}_{\alpha\beta}$ and $\mathbf{v}_1^\alpha$ is necessary (we will come back to this point). Second, equation (10) is solved for $\rho^{(2)}_{\alpha\beta}$ in terms of $\mathbf{v}_1^\alpha$. Third, this solution is put in to (8), leading to an equation for the motion of any particle of species α alone. Due to the stochastic nature of the previous equations, the complete procedure can be carried out by applying the Medina–Noyolas's contraction procedure of non–Markov processes [8], which leads to the generalized Langevin equation:

$$
m_\alpha \frac{d\mathbf{v}^\alpha}{dt} = -\zeta_\alpha \mathbf{v}^\alpha - \int_0^t \Delta\zeta_\alpha(t - t') \mathbf{v}^\alpha(t') dt' + \mathbf{F}^\alpha + \mathbf{f}^\alpha.
\tag{11}
$$

Here, $\mathbf{F}^\alpha(t)$ is a colored random force related to $\Delta\zeta_\alpha(t)$ through a FDT. The contraction also results in explicit expressions for $\Delta\zeta_\alpha(t)$, as shown down.

Equation (11) refers only to one particle of species α explicitly. The rest of the particles of the same species were contracted, as well as all particles of the other species. Therefore, equation (11) represents an effective one particle dynamics, where the rest of the systems is included only in the friction coefficient ζ_α (the water molecules) and in its increment $\Delta\zeta_\alpha(t)$ (the macroions, counterions and salt ions). In the following section we apply this scheme in order to calculate the long–time self-diffusion coefficient of polystyrene spheres.

3.   LONG–TIME SELF–DIFFUSION COEFFICIENT

We can take the bulk value $n_\alpha = N_\alpha/V$ for $\rho_\alpha^{(1)}(\mathbf{r}_1; t)$ and rewrite the quantity $\rho_{\alpha\beta}^{(2)}(\mathbf{r}_1, \mathbf{r}_2; t) \equiv \rho_\alpha^{(1)}(\mathbf{r}_1; t)\rho_{\alpha\beta}(\mathbf{r}_2/\mathbf{r}_1; t)$ as $n_\alpha n_\beta(\mathbf{r}_{12}; t)$, where $\rho(\mathbf{r}_2/\mathbf{r}_1; t)$ is the conditional reduced distribution function of finding particle 2 of species β in $\mathbf{r}_2$ at time t, given the particle 1 of species α is in $\mathbf{r}_1$ at the same time. Also, $n_\beta(\mathbf{r}_{12}; t) = n_\beta^{eq}(\mathbf{r}_{12}) + \delta n_\beta(\mathbf{r}_{12}; t)$ is the instantaneous local density of particles of species β around a central particle of species α with equilibrium value $n_\beta^{eq}(\mathbf{r}_{12})$ and fluctuations $\delta n_\beta(\mathbf{r}_{12}; t)$ around the equilibrium state. Taking into account these definitions, equation (8) reads

$$m_\alpha \frac{d\mathbf{v}^\alpha}{dt} = -\zeta_\alpha \mathbf{v}^\alpha + \sum_\beta \int (\nabla u_{\alpha\beta}(r)) \, \delta n_\beta(\mathbf{r}; t) d\mathbf{r} + \mathbf{f}^\alpha. \qquad (12)$$

We take for $\partial \rho_{\alpha\beta}^{(2)}/\partial t$ one of the simplest approaches available in the literature; the modified Fick's approximation [8]:

$$\frac{\partial \delta n_\beta}{\partial t} = (D_\alpha + D_\beta) \, \nabla \cdot n_\beta^{eq}(\mathbf{r}) \nabla \sum_\gamma \int \frac{\delta n_\gamma(\mathbf{r}'; t)}{\langle \delta n_\beta(\mathbf{r}) \delta n_\gamma(\mathbf{r}') \rangle^{eq}} d\mathbf{r}', \qquad (13)$$

where $\langle \delta n_\beta(\mathbf{r}) \delta n_\gamma(\mathbf{r}') \rangle^{eq}$ means the equilibrium density fluctuations correlation function. Equation (13) is, ad hoc, linear in δn_γ and represents a particular case of the general equation (9). We also linearize equation (10) in the small quantities $\mathbf{v}^\beta$ and δn_β;

$$\frac{d\delta n_\beta}{dt} = \mathbf{v}^\beta \cdot \nabla n_\beta^{eq} + \frac{\partial \delta n_\beta}{\partial t}. \qquad (14)$$

After contracting our equations following references [8] and [9] in order to generate an effective dynamics for a polystyrene sphere, we get

$$\begin{aligned}
\Delta \zeta_m &= \int_0^\infty \Delta \zeta_m(t) dt \\
&= \frac{\zeta_m}{6\pi^2} \sum_{\alpha=m,c,+,-} \frac{n_\alpha}{(1 + D_\alpha/D_m)} \int_0^\infty h_{m\alpha}^2(q) q^2 dq. \qquad (15)
\end{aligned}$$

Here, $h_{m\alpha}(q)$ is the Fourier transform of $h_{m\alpha}(r)$.

Once $\Delta \zeta_m$ is determined, de long–time self–diffusion coefficient D_S^L of polystyrene particles follows immediately from the velocity self–correlation function $\langle \mathbf{v}^m(t) \cdot \mathbf{v}^m(0) \rangle / 3$, because its time integral is related to the mean–square displacement. Therefore, one gets

$$D_S^L = \frac{k_B T}{\zeta_m + \Delta\zeta_m} \tag{16}$$

or

$$\frac{D_S^L}{D_m} = \left[1 + \frac{n_m}{12\pi^2} \int_0^\infty h_{mm}^2(q) q^2 dq \right.$$
$$\left. + \frac{1}{6\pi^2} \sum_{\alpha=c,+,-} \frac{n_\alpha}{(1 + D_\alpha/D_m)} \int_0^\infty h_{m\alpha}^2(q) q^2 dq \right]^{-1}, \tag{17}$$

with substitution of (15) into (16). The second term in the parenthesis accounts for self–friction (SF, the friction due to the interaction between the tracer particle and the other macroions). In order to evaluate it, we numerically calculate $h_{mm}(r)$ within the OCM and the thermodynamically self–consistent Rogers–Young (RY) approximation [10]

$$g_{mm}(r) = \exp\left[-\beta u_{mm}(r)\right] \left[1 + \frac{\exp\left[(h_{mm}(r) - c_{mm}(r)) f(r)\right] - 1}{f(r)} \right]. \tag{18}$$

Here, the value of the mixture parameter α in $f(r) = 1 - e^{-\alpha r}$ is fixed by demanding the equality of the isothermal compressibility calculated from the compressibility equation of state and from the virial equation of state [10, 11]. The third term in the parenthesis of equation (17) accounts for electrolyte friction (EF, the friction due to the interaction between the tracer particle and the counterions and salt ions). We evaluate it within the PM and the Debye–Hückel limit;

$$h_{m\alpha}(r) = \left[1 + \beta \frac{Q_m Q_\alpha}{\epsilon} \frac{e^{\kappa a_m}}{1 + \kappa a_m} \frac{e^{-\kappa r}}{r} \right] \Theta(r - a_m) - 1, \tag{19}$$

which considers point–like counterions and salt ions. Here, $\Theta(r - a_m)$ is the Heaviside's step function. Finally, D_S^L/D_m remains

$$\frac{D_S^L}{D_m} = \left[1 + \frac{n_m}{12\pi^2} \int_0^\infty h_{mm}^2(q) q^2 dq + K Q_m^2 \sum_{\alpha=c,+,-} \frac{n_\alpha Q_\alpha^2}{1 + a_m/a_\alpha} \right]^{-1}, \tag{20}$$

with

$$K = \left[3 \left(k_B T \epsilon\right)^2 \kappa \left(1 + \kappa a_m\right)^2 / 2\pi \right]^{-1}. \tag{21}$$

We now use equation (20) in order to evaluate the long–time self–diffusion coefficient in systems with different volume fractions $\varphi_m =$

$4\pi n_m a_m^3/3$ of polystyrene spheres and salt concentrations n_s, keeping $Q_m = 500e^-$ and $a_m = 50$nm constant. For the small ions we take $Q_c = Q_+ = -Q_- = e^+$ and the radii $a_c = a_+ = 0.18$nm and $a_- = 0.12$nm, obtained from the value of the ionic mobility of Na^+ and Cl^- in dilute aqueous solution by means of the Stokes friction. We also take $T = 300$K and $\epsilon = 78.2$. The results are shown in Figure 1. Let us first consider salt concentration effects, i.e., the case of only one macroion immersed in an electrolyte solution ($\varphi_m = 0$). Electroneutrality leads to $n_c = 0$ and $n_+ = n_- = n_s$. For very low salt contents the diffusion coefficient is the corresponding to a free particle. By increasing n_s the diffusion coefficient is reduced until a minimum is reached at $\kappa a_m = 1$. Then, D_S^L increases again, by increasing n_s, towards D_m. The curve looks like an inverted bell. For macroions concentration effects we have that the electroneu-

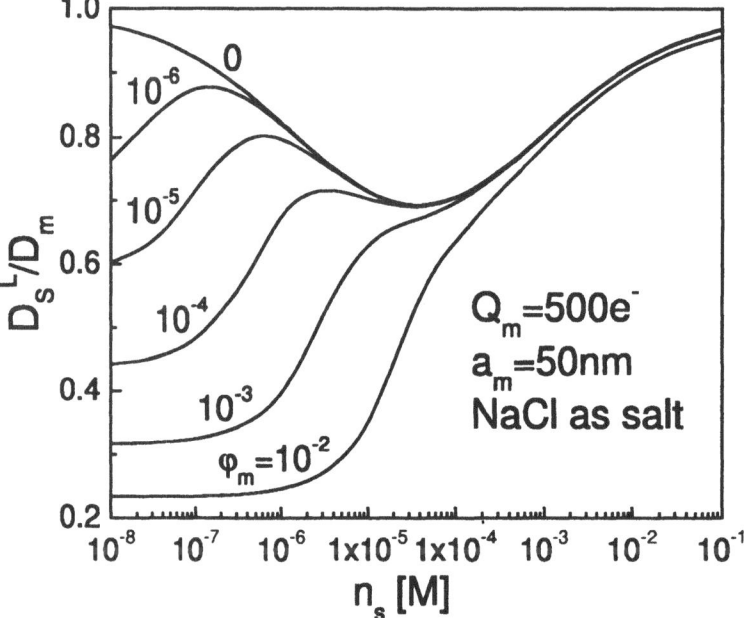

Figure 1.   The figure shows macroions and salt concentration effects on the reduced long–time self–diffusion coefficient of macroions with charge $500e^-$ and radius 50nm. The used salt was NaCl.

trality leads to $n_c = 500n_m$ and $n_+ = n_- = n_s$. We find in the region with $\kappa a_m \gg 1$ that the screening of the interaction between macroions is so large that they behave like isolated particles in an electrolyte solution. In the region with $\kappa a_m \ll 1$ however the macroion–macroion interaction is decisive; the diffusion coefficient decreases with φ_m. From the scale of Figure 1, it becomes clear that D_S^L decreases as $\ln \varphi_m$, which agrees qualitatively with the available experimental measurements [12]. Moreover, in the transition region ($\kappa a_m \approx 1$) a non–monotonic behavior of D_S^L as function of n_s is found for $\varphi_m \lesssim 10^{-4}$. This is a consequence of the competition between self-friction and electrolyte friction, which has been recently observed [12]. For $\varphi_m \gtrsim 10^{-3}$ the diffusion coefficient is always an increasing function of n_s.

4. ELECTROPHORETIC MOBILITY

Let us turn to the description of the motion of a macroion under the influence of an external constant electric field $\mathbf{E}$. Equation (6) and (7) must be modified by adding the terms $Q_\alpha \mathbf{E}$ and $-\beta \left(Q_\beta \mathbf{E} \right) P^N$ to their right hands, respectively [7]. It results in the terms $Q_\alpha \mathbf{E}$ and $-\beta D_\alpha Q_\alpha \mathbf{E} \cdot \nabla_1 \rho_{\alpha\beta}^{(2)} - \beta D_\beta Q_\beta \mathbf{E} \cdot \nabla_2 \rho_{\alpha\beta}^{(2)}$, which must be in turn added to the right hands of equations (8) and (9), respectively. The contraction of the modified equations leads to the generalized Langevin equation in the external electric field $\mathbf{E}$:

$$
\begin{aligned}
m_\alpha \frac{d\mathbf{v}^\alpha}{dt} = & -\zeta_\alpha \mathbf{v}^\alpha - \int_0^t \Delta\zeta_\alpha(t-t')\mathbf{v}^\alpha(t')dt' + Q_\alpha \mathbf{E} \\
& + \int_0^t \Delta Q_\alpha(t-t')\mathbf{E}dt' + \mathbf{F}^\alpha + \mathbf{f}^\alpha.
\end{aligned}
\tag{22}
$$

The electrophoretic mobility of polystyrene spheres μ^m follows from the steady state average of this equation;

$$
\frac{\mu^m}{\mu_0^m} = \frac{1 + \Delta Q_m/Q_m}{1 + \Delta\zeta_m/\zeta_m},
\tag{23}
$$

with $\Delta\zeta_m$ (ΔQ_m) given by the time integral of $\Delta\zeta_m(t)$ ($\Delta Q_m(t)$). The quantity μ_0^m stands for the free particle mobility Q_m/ζ_m.

Making use of the bulk value $n_\alpha = N_\alpha/V$ for $\rho_\alpha^{(1)}(\mathbf{r}_1; t)$ and rewriting the quantity $\rho_{\alpha\beta}^{(2)}(\mathbf{r}_1, \mathbf{r}_2; t) \equiv \rho_\alpha^{(1)}(\mathbf{r}_1; t)\rho_{\alpha\beta}(\mathbf{r}_2/\mathbf{r}_1; t)$ as $n_\alpha n_\beta(\mathbf{r}_{12}; t)$, but this time with $n_\beta(\mathbf{r}_{12}; t) = n_\beta^{SS}(\mathbf{r}_{12}) + \delta n_\beta(\mathbf{r}_{12}; t)$ the instantaneous local density of particles of species β around a central particle of species α with steady state value $n_\beta^{SS}(\mathbf{r}_{12})$ and fluctuations $\delta n_\beta(\mathbf{r}_{12}; t)$ around the steady state, equation (12) is modified by adding the term $Q_\alpha \mathbf{E}$ to its right hand. Moreover, the appropriate extension of the modified Fick's approximation succeeds by changing the superscript equilibrium (eq) by steady state (ss), and by adding the term $-\beta \left(D_\alpha + D_\beta \right) Q_\beta \mathbf{E} \cdot \nabla n_\beta^{SS}(\mathbf{r})$ to the right hand of equation (13). After contracting our new equations following references [8] and [13] in order to generate an effective dynamics

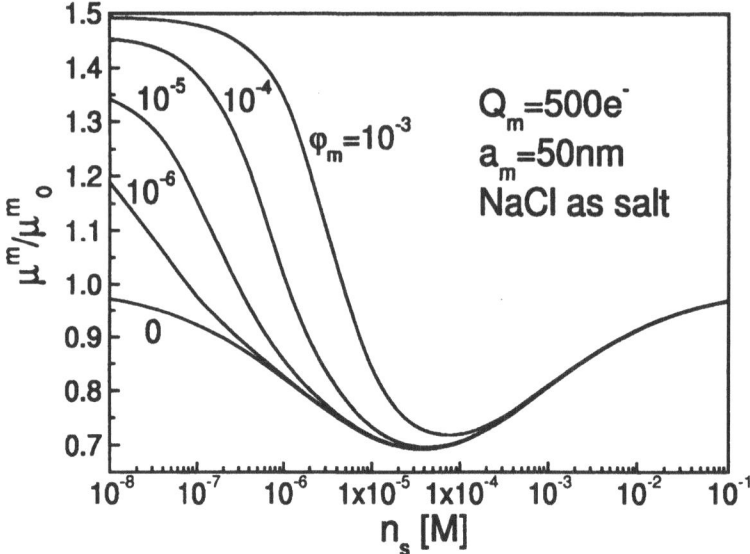

Figure 2.   The figure shows macroions and salt concentration effects on the reduced electrophoretic mobility of macroions with charge $500e^-$ and radius 50nm. The used salt was NaCl.

for a polystyrene sphere, we get

$$\frac{\mu^m}{\mu_0^m} = \frac{1 + I_{mm} + \sum_{\alpha=c,+,-} Q_\alpha I_{m\alpha}/Q_m}{1 + I_{mm}/2 + \sum_{\alpha=c,+,-} I_{m\alpha}/\left(1 + D_\alpha/D_m\right)}, \qquad (24)$$

with

$$I_{\alpha\beta} = \frac{n_\beta}{6\pi^2} \int_0^\infty h_{\alpha\beta}^2(q) q^2 dq. \qquad (25)$$

The friction increment is the same as in the previous section, but this time we also have an expression for the electrokinetic charge. In the derivation of (24) we assumed that the statistical averages in the stationary state can be taken as in equilibrium, which is expected to be valid for weak electric fields. By the way, we also assumed an homogeneous fluid. Making the same approximations for the structure functions as in the previous section, we get

$$\frac{\mu^m}{\mu_0^m} = \frac{1 + I_{mm} + K Q_m \sum_{\alpha=c,+,-} n_\alpha Q_\alpha^3}{1 + I_{mm}/2 + K Q_m^2 \sum_{\alpha=c,+,-} n_\alpha Q_\alpha^2/\left(1 + a_m/a_\alpha\right)}, \qquad (26)$$

with K given by (21). Let us now evaluate the electrophoretic mobility of polystyrene spheres for the same parameters as in Figure 1. The results are shown in Figure 2. Let us first consider salt concentration effects, i.e., the case of only one macroion immersed in an electrolyte solution

($\varphi_m = 0$). Electroneutrality leads to $n_c = 0$ and $n_+ = n_- = n_s$. For very low salt contents the mobility is the corresponding to a free particle. By increasing n_s, the mobility is reduced until a minimum is reached at $\kappa a_m = 1$ and then μ^m increases again towards μ_0^m. The curve looks like an inverted bell. Actually, it coincides with that for D_S^L, because the electrokinetic charge is zero for symmetric salt and $\varphi_m = 0$.

Regarding the response of the electrophoretic mobility to the macroion concentration we first have that the electroneutrality leads to $n_c = 500 n_m$ and $n_+ = n_- = n_s$. We find in the region with $\kappa a_m \gg 1$ that the screening of the interaction between macroions is so large that they behave like isolated particles in an electrolyte solution. In the region with $\kappa a_m \ll 1$ however the macroion–macroion interaction is decisive; the mobility increases with φ_m. From the scale of Figure 2, it becomes clear that μ^m increases as $\ln \varphi_m$, which agrees with the available experimental measurements [14]. Moreover, the inverted bell becomes more asymmetric due to this enhancement, and its minimum simultaneously moves to larger values of κa_m. Its new position is roughly given by the value of n_s where the curves for $\varphi_m \neq 0$ meet the one for $\varphi_m = 0$. ¿From a visual inspection of Figure 2, it seems to occur when n_s[in moles/dm^3] $\approx \varphi_m$ (note that this criterion is valid for the magnitudes, but not for the unities). Surprisingly, it is exactly what Weber observed in his experiments [15]. We also got similar results for other values of Q_m. In particular, the position of the minimum does not appreciably depend on the value of Q_m. This explains why we can compare with the results of Weber without to know anything about the charge on his particles.

5. DISCUSSION AND CONCLUSIONS

We presented here a general method for constructing effective one particle dynamics in colloidal systems, which results from a contraction of the description of multicomponent systems based on the many particles Langevin and Smoluchowski equations. Its applicability is illustrated by calculating the long-time self-diffusion coefficient of charged particles, as well as its electrophoretic mobility. The comparison of our results with available experimental data confirms that our theoretical scheme captures macroion and salt concentration effects with qualitative accuracy. This is however a great goal, because some of the explained phenomena were captured here for the first time in a theory. In particular, the shift of the minimum of μ^m, as function of salt content, with increasing macroion concentration and the simultaneous enhancement of μ^m.

Although not shown in this paper, our theoretical scheme can also accounts for other new phenomena, like the saturation of μ^m for very structured systems [14] and the macroion charge inversion in presence of asymmetric salt [16]. These results will be published elsewhere [17].

Acknowledgments

Very helpful discussions with M. Medina–Noyola and R. Weber are gratefully acknowledged. This work was supported by CONACyT–Mexico (grants 33815–E and 1748P–E).

References

[1] R. Castañeda–Priego, A. Rodríguez–López, M. D. Carbajal–Tinoco, P. González–Mozuelos and J. M. Méndez–Alcaraz, Developments in Mathematical and Experimental Physics, Volume B: Statistical Physics and Beyond, edited by Alfredo Macias, Enrique Díaz, Francisco J. Uribe, (Kluwer Academic/Plenum Press 2002).

[2] J. M. Méndez–Alcaraz, M. Chávez–Páez, B. D'Aguanno and R. Klein, Physica **A220** (1995) 173.

[3] P. González–Mozuelos and M. D. Carbajal–Tinoco, J. Chem. Phys. **109** (1998) 11074.

[4] M. Medina–Noyola and D. A. McQuarrie, J. Chem. Phys. **73** (1980) 6279.

[5] J. P. Hansen and I. R. McDonald, *Theory of Simple Liquids* (Academic Press, London, 1986).

[6] J. M. Méndez–Alcaraz and R. Klein, Phys. Rev. **E61** (2000) 4095.

[7] T. J. Murphy and J. L. Aguirre, J. Chem. Phys. **57** (1972) 2098.

[8] M. Medina–Noyola, Faraday Discuss. Chem. Soc. **83** (1987) 21.

[9] J. M. Méndez–Alcaraz and O. Alarcón–Waess, Physica **A268** (1999) 75.

[10] F. J. Rogers and D. A. Young, Phys. Rev. **A30** (1984) 999.

[11] J. M. Méndez–Alcaraz, B. D'Aguanno and R. Klein, Langmuir **8** (1992) 2913.

[12] D. Rudhardt and P. Leiderer, private communications.

[13] O. Alarcón–Waess and R. M. Velasco, J. Chem. Phys. **101** (1994) 5110.

[14] M. Evers, N. Garbow, D. Hessinger and T. Palberg, Phys. Rev. **E57** (1998) 6774.

[15] M. Deggelmann, T. Palberg, M. Hagenbüchle, E. E. Maier, R. Krause, C. Graf and R. Weber, J. Colloid Interface Sci. **143** (1991) 318.

[16] M. Lozada–Cassou and E. González–Tovar, J. Colloid Interface Sci. **239** (2001) 285.

[17] J. M. Méndez–Alcaraz and O. Alarcón–Waess, manuscript in preparation.

ROTATING UNSTABLE LANGEVIN–TYPE DYNAMICS AND NONLINEAR EFFECTS

J. I. Jiménez–Aquino, M. Romero–Bastida

Departamento de Física, Universidad Autónoma Metropolitana, Iztapalapa

Apdo. Postal 55-534, México, D. F. 09340 México

ines@xanum.uam.mx

Abstract   In this work we propose how to characterize nonlinear rotating unstable systems with a Langevin–type dynamics in the presence of a constant external force, through the Passage Time Distribution (PTD). Although the rotating Langevin–type equation is proposed in two dynamical representations **x** and **y**, being **y** the transformed space of coordinates by means of a time-dependent rotation matrix, the study is basically given in the space of coordinates **y**. A laser system in the presence of an external injected signal is considered as a prototype system, which admits a description in terms of two variables.

Keywords: Stochastic process, passage time distribution, laser field.

1.   INTRODUCTION

Very recently a general characterization of linear rotating unstable systems through the Passage Time Distribution (PTD) has been proposed in a theoretical scheme of the Langevin–type dynamics [1]. For rotating unstable systems we mean those which, once leaving the initial unstable state, describe practically deterministic rotational trajectories to reach the steady state or some approximation of it. As it is well known, the PTD is essentially defined as the mean time at which the system reaches a prescribed reference value, after it leaves the initial unstable state as a consequence of stochastic fluctuations.

In the later eighties, another time scale called Non–Linear Relaxation Times (NLRT) was also proposed as an appropriate quantity in the complete dynamical characterization of the decay of unstable steady states, taking into account the dynamical evolution evolution of the system toward its stable steady state. The theory was established in the context of the Fokker–Planck formulation with their corresponding mathemati-

Developments in Mathematical and Experimental Physics,
Volume B: Statistical Physics and Beyond
Edited by Macias *et al.*, Kluwer Academic/Plenum Publishers, 2003

57

cal difficulties, principally in the description of the problem with colored noise [2].

The time characterization of nonlinear unstable systems through the NLRT in connection with another well known approach called the Quasi–Deterministic (QD) approach [3, 4, 5] was really proposed in Ref. [6], in the context of Langevin–type dynamics which has shown to be much easier than the Fokker–Planck point of view. Following the ideas of references [1] and [6], we propose in this work how to deal with the non-linearities in the decay of rotating unstable systems, through the PTD. The rotating Langevin dynamics is proposed in two dynamical representations $\mathbf{x}$ and $\mathbf{y}$, where $\mathbf{y}$ is the transformed space of coordinates by means of a time–dependent rotation matrix. The time characterization is given in the transformed space of coordinates because in this space, the set of variables $\mathbf{y}$ are not coupled and therefore the Langevin associated systematic force is derived from a potential. The theory is applied to study the decay of a laser system in presence of a large external electric field, taking into account the saturation effects.

2.   THE ROTATING UNSTABLE LANGEVIN DYNAMICS

The rotating unstable Langevin type equation for the column vector $\mathbf{x}$ in presence of constant external force can be written as

$$\dot{\mathbf{x}} = a\mathbf{x} + W\mathbf{x} + n(r)\mathbf{x} + \mathbf{f}_e + \mathbf{z}(t), \tag{1}$$

where a is positive, W is a $n \times n$ real antisymmetric matrix such that $W^T = -W$ and W^T its transposed. The scalar function $n(r)$ accounts for nonlinear contributions due to the fact that $r \equiv x^2 = \mathbf{x}^T\mathbf{x}$, being r the square of the norm of the vector and $\mathbf{z}(t)$ the fluctuating force whose elements $\xi_i(t)$ satisfy the property of Gaussian white noise with zero mean value and correlation function

$$\langle \xi_i(t)\xi_j(t') \rangle = 2Q_{ij}\,\delta_{ij}\,\delta(t - t'), \tag{2}$$

where Q_{ij} is the matrix which represents the noise intensity. The linear systematic force $\mathbf{f}_s = a\mathbf{x} + W\mathbf{x}$ is not in general derived from a potential, because $\nabla \times \mathbf{f}_s = \nabla \times W\mathbf{x} \neq 0$ and therefore the rotating character of the dynamics (1) is due to the properties of the matix W.

A different equivalent Langevin type dynamics can be established if we make the change of variable $\mathbf{y} = e^{-Wt}\mathbf{x}$, such that

$$\dot{\mathbf{y}} = a\mathbf{y} + n(r)\mathbf{y} + \Re(t)^{-1}\mathbf{f}_e + \Re(t)^{-1}\mathbf{z}(t), \tag{3}$$

where the factor $e^{Wt} = \Re(t)$ is in general a time dependent orthogonal rotation matrix [8], which satisfies that $\Re^T(t) = \Re^{-1}(t)$ and therefore

$e^{-Wt} = \Re^{-1}(t)$. $n(r)$ is the same function because r is invariant, i.e., $r \equiv x^T x = y^T y$. Clearly in this dynamics, the external force and the internal noise are rotational.

To take into account the nonlinearities in the time characterization of the dynamics (3), we first require of the definition of the deterministic time evolution of the r variable as proposed in Ref. [6], that is

$$\dot{r} = f(r) = \frac{r(r_{st} - r)}{C_0 + rg(r)} \, , \tag{4}$$

where $f(r)$ is a general unstable potential function, $C_0 = r_{st}/2a$ and $g(r) > 0$ is a polynomial. The function $f(r)$ has two roots, one is at $r = 0$ which is the unstable state and the other root is at $r = r_{st}$, which corresponds to the stable steady state. Eq. (4) is compatible with the deterministic part of (3) according to explicit form of $n(r)$.

From definition (4) we can establish that the random passage time at which the system reaches a reference value $r_R = R^2$, from the initial condition $r(0)$ can be defined as

$$t = \int_{h^2(t)}^{R^2} \frac{dr}{f(r)} \, , \tag{5}$$

where the initial condition $r(0)$ has been substituted by the stochastic process $h^2(t)$ for not so large times, in order to take into account the rotational effects of the dynamics (3). So, substituting Eq. (4) into Eq. (5) we get the random passage time

$$t = \frac{1}{2a} \ln \left[\frac{r_{st}}{h^2(t)} \right] + \frac{1}{2a} \ln \left[\frac{\gamma}{1 - \gamma} \right] + \frac{1}{2a} \ln \left[1 - \frac{h^2(t)}{r_{st}} \right] + \int_{h^2(t)}^{R^2} \frac{g(r)}{r_{st} - r} \, dr \, , \tag{6}$$

where the constant $\gamma = r_R/r_{st}$. Evidently $g(r)$ is a function which depends on the type of model. The first term of Eq. (6) comes from an analysis of the linear approximation of Eq. (3), whose solution can be obtained according to the analysis of Ref. [1]. In this case it can be proved that PTD can be approximated by

$$\langle t \rangle \equiv t_P = \frac{1}{2a} \ln \left(\frac{r_{st}}{|\langle \mathbf{h}(t_P) \rangle|^2} \right) + \frac{1}{2a} \ln \left[\frac{\gamma}{1 - \gamma} \right] + \frac{1}{2a} \frac{|\langle \mathbf{h}(t_P) \rangle|^2}{r_{st}}$$
$$+ \left\langle \int_{h^2(t)}^{R^2} \frac{g(r)}{r_{st} - r} \, dr \right\rangle , \tag{7}$$

where $|\langle \mathbf{h}(t_P) \rangle|^2 = \sum_i \langle h_i(t_P) \rangle^2$ and the statistical properties of the process $\mathbf{h}(t)$ are given by

$$\langle h_i(t) \rangle = \int_0^t e^{-as} \Re_{ki}(s) f_{e_k} ds \,, \tag{8}$$

and

$$\langle h_i(t) h_j(t) \rangle = \langle h_i(t) \rangle \langle h_j(t) \rangle + \frac{Q}{a}(1 - e^{-2at}) \, \delta_{ij} \,. \tag{9}$$

2.1.   THE LASER SYSTEM

The Langevin–type equation for the dimensionless laser field $E = E_1 + iE_2$ of a single–mode in the presence of a constant external injected signal in which we are interested reads as [4, 5, 7]

$$\dot{E} = (-kE + if)E + \frac{F}{1 + (A/F)I}E + k_e E_e + \xi(t) \,, \tag{10}$$

where $\xi(t)$ is the spontaneous emission Gaussian noise of zero mean and correlation

$$< \xi(t)\xi^*(t') > = 2\epsilon\delta(t - t') \,, \tag{11}$$

k is the cavity decay rate in s^{-1}, f is the detuning parameter between the laser field and the injected signal, F is the gain parameter(s^{-1}), A the saturation parameter(s^{-1}), $I = |E|^2 = E_1^2 + E_2^2$ the intensity of the laser field, k_e is the coupling parameter between the injection field E_e and the laser filed and $\xi(t)$ is the internal noise with strength $\epsilon(s^{-1})$. Both k_e and E_e are taken as real numbers.

A third order approximation in the electric field, the dynamics (10) reduces to

$$\dot{E} = aE + ifE - A|E|^2 E + k_e E_e + \xi(t) \,, \tag{12}$$

where $a = F - k$. For this dynamics the steady state value of the intensity is then $I_{st} = a/A$. The dynamics (10) is quite compatible with that given in (1) or (3) in the case of two variables. In this case the parameter $\omega = f$; the real part of the external field is $f_{e_1} = k_e E_e$ whereas the imaginary part $f_{e_2} = 0$. We use the same experimental values used in Ref. [5] to show a single stochastic trajectory of the system on the plane (y_1, y_2) which describes "loops", as can be clearly seen in Fig. 1.

The function $g(r)$ in this case is identically zero. Then we can show that, for large amplitude of the external field, the PTD taking into account the nonlinear effects can be written as

$$t_P = t_0 - \frac{1}{2a} \ln[1 + \phi(t_P)] + \frac{1}{2aI_{st}} \frac{k_e^2 E_e^2}{(a^2 + \omega^2)}[1 + \phi(t_P)] \,, \tag{13}$$

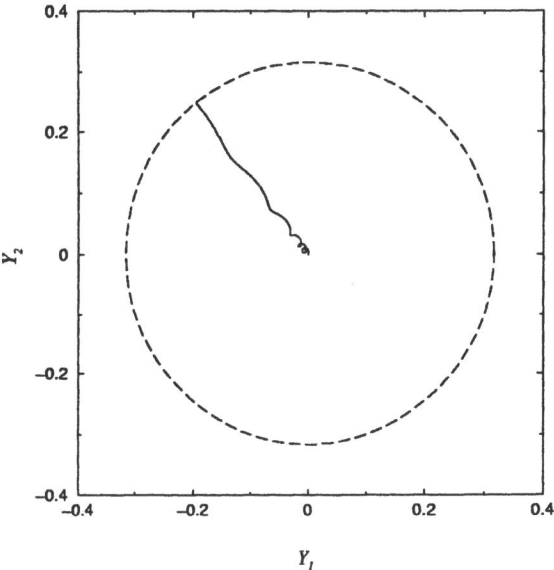

Figure 1. Dynamical evolution of a single stochastic trajectory in its way to reach the circle of radius $R^2 = 0.1$ for the two–variable system defined by Eq. (12) written in the form (3).

where

$$t_0 = \frac{1}{2a} \ln \left[\frac{I_{st}(a^2 + \omega^2)}{k_e^2 E_e^2} \right] + \frac{1}{2a} \ln \left[\frac{\gamma}{1 - \gamma} \right], \qquad (14)$$

and $\phi(t) = [e^{-2at} - 2e^{-at} \cos \omega t]$. The time scale (13) can be calculated through the iterative procedure

$$t_P^{(0)} = t_0,$$

$$t_P^{(n+1)} = t_0 - \frac{1}{2a} \ln[1 + \phi(t_P^{(n)})] + \frac{1}{2aI_{st}} \frac{k_e^2 E_e^2}{(a^2 + \omega^2)} [1 + \phi(t_P^{(n)}))] \quad (15)$$

In future works our proposal will be developed with more detail and the results will be compared with numerical simulation.

Acknowledgments

The authors wish to thank to the organizers of the "First Mexican Meeting on Mathematical and Experimental Physics" held at El Colegio Nacional from 10–14 September, 2001, for the kind invitation. Financial support from Consejo Nacional de Ciencia y Tecnología (CONACYT) México is also acknowledged.

References

[1] J. I. Jiménez–Aquino and M. Romero–Bastida, *Phys. Rev.* **E64** (2001) 050102.

[2] J. Casademunt, J. I. Jiménez–Aquino, and J. M.Sancho, *Phys. Rev.* **A40** (1989) 5905.

[3] F. de Pascuale, J. M. Sancho, M. San Miguel and P. Tartaglia, *Phys. Rev. Lett.* **56** (1986) 2473.

[4] S. Balle, F. de Pasquale and M. San Miguel, *Phys. Rev.* **A41** (1990) 5012.

[5] J. Dellunde, J. M. Sancho and M. San Miguel, *Opt. Comm* **39** (1994) 435.

[6] J. I. Jiménez–Aquino and J. M. Sancho, *Phys. Rev.* **E47** (1993) 1558.

[7] G. Vemuri and R. Roy, *Optics. Comm.* **77** (1990) 318.

[8] E. Piña, *Dinámica de Rotaciones*, (Universidad Autónoma Metropolitana, México, 1996) (in Spanish).

MESOSCOPIC THERMODYNAMICS FOR BROWNIAN PARTICLES

M. Mayorga
Facultad de Ciencias
Universidad Autónoma del Estado de México
Av. Instituto Literario 100, Toluca, CP. 50000, México.

Abstract   We review the main statements of the mesoscopic irreversible thermodynamics. Afterwards we show how the approach is appropriate to
construct the balance equations of interacting Brownian particles in solution. The application of the scheme to the case when the particles are
in a temperature gradient, permit to obtain the connection between the
collective diffusion and the thermal diffusion coefficient in terms of the
pair interaction potential between the particles, and their corresponding
pair correlation function.

Keywords: Non–equilibrium thermodynamics, colloids, collective diffusion, thermal
diffusion.

1.   INTRODUCTION

In this introduction we intend to review a recent mesoscopic approach
introduced by P. Mazur , where the treatment of thermodynamic fluctuations are incorporated in the macroscopic framework of non–equilibrium
thermodynamics [1]. As Mazur quote, the term mesoscopic is used in
the original sense introduced by van Kampen, to mean "...the stochastic
description in terms of macroscopic variables.... It comprises both the
deterministic laws and the fluctuations about them". On the other hand
as is well known, the central quantity in the description of macroscopic
non–equilibrium thermodynamics is the entropy S of the system. If we
are interested for example, in the study of a closed adiabatically insu-
lated system, the macroscopic state is described by a set of n independent
variables A_i $(i = 1, ..., n)$ corresponding to extensive properties, masses,
energies, etc. Here, in addition we take into account the conservation
laws, for small subsystems large enough to make a thermodynamic description possible. An arbitrary state may be specified by deviations α_i

Developments in Mathematical and Experimental Physics,
Volume B: Statistical Physics and Beyond
Edited by Macias *et al.*, Kluwer Academic/Plenum Publishers, 2003

from the equilibrium state

$$\alpha_i = A_i - A_i^0, \tag{1}$$

where A_0 stands for the equilibrium value of the variables A_i.

The total entropy is a time functional through the state variables, namely,

$$S(t) = S(\alpha_1(t), ..., \alpha_n(t)), \tag{2}$$

here $\vec{\alpha} = (\alpha_1, ..., \alpha_n)$ characterizes the macroscopic state as a point in an n–dimensional cartesian phase space. From the entropy (2) one can determine, the probability density distribution function of the equilibrium fluctuations of the state variables. In addition, such distribution function plays an important role in the Onsager's analysis of the connection between fluctuations and irreversible processes .

On the other hand, in order to incorporate the fluctuations of macroscopic quantities we need to describe the entropy as a mesoscopic quantity, namely,

$$S\left(\left\{P\left(\vec{\alpha},t\right)\right\}\right) = -k_B \int P\left(\vec{\alpha},t\right) \ln\left(\frac{P\left(\vec{\alpha},t\right)}{P_0\left(\vec{\alpha},t\right)}\right) d\vec{\alpha} + S_{eq}, \tag{3}$$

where $P\left(\vec{\alpha},t\right)$ is the probability that the system is at time t in the state $\vec{\alpha}$, S_{eq} is the total equilibrium entropy which as is usual in thermodynamics, is a function of the total energy, the volume and the total mass, i.e., $S\left(\left\{P\left(\vec{\alpha},t\right)\right\}\right)$ is itself a macroscopic function of state. On the other hand, the quantity which accounts for the irreversible processes in the system, i.e. the entropy production, can be obtained with the evolution equation of the functional $S\left(P\left(\vec{\alpha},t\right)\right)$. This last item can be reached with the assumption that the probability obeys a continuity equation in the phase space, namely,

$$\frac{\partial P}{\partial t} = -\sum_{i=1}^{N} \frac{\partial}{\partial \alpha_i} J_{u_i}, \tag{4}$$

where J_{u_i} is the flux in the phase space. In order to obtain the flux J_{u_i}, we apply the equation (4) to the entropy functional (3) and we obtain the next relation,

$$\frac{dS}{dt} = \int \sigma\left(\vec{\alpha},t\right) d\vec{\alpha} \tag{5}$$

where we demand $\sigma\left(\vec{\alpha}, t\right) \geq 0$, in addition, the entropy production has the next byproduct form,

$$\sigma\left(\vec{\alpha}, t\right) = -k\sum_{i=1}^{N} J_{u_i}\frac{\partial}{\partial\alpha_i}\ln\left(\frac{P\left(\vec{\alpha}, t\right)}{P_0\left(\vec{\alpha}, t\right)}\right). \tag{6}$$

Here we can follow the scheme of non-equilibrium thermodynamics [2], and afterwards of identify the fluxes J_{u_i} and forces $\frac{\partial}{\partial\alpha_i}\ln\left(\frac{P\left(\vec{\alpha}, t\right)}{P_0\left(\vec{\alpha}, t\right)}\right)$ we establish a linear relationship between them,

$$J_{u_i} = k\sum_{j=1}^{N} L_{ij}\frac{\partial}{\partial\alpha_i}\ln\left(\frac{P\left(\vec{\alpha}, t\right)}{P_0\left(\vec{\alpha}, t\right)}\right) \tag{7}$$

where L_{ij} represents the phenomenological Onsager's coefficients and have a reciprocity relationship and in general depend of $\vec{\alpha_i}$. With the definition of the next coefficient

$$\beta_{ij} = \frac{mL_{u_iu_j}}{PT_{eq}}, \tag{8}$$

we can rewrite the fluxes

$$J_{u_i} = -\sum_{\substack{j\neq i \\ j=1}}^{N} \beta_{ij}\left(P\alpha_i + \frac{kT_{eq}}{m}\frac{\partial P}{\partial\alpha_i}\right). \tag{9}$$

According to this representation for the fluxes, the continuity equation (4) now is a Fokker–Planck equation for the probability P of the fluctuations α_i, and this means that such fluctuations represent a Gaussian–Markov stochastic process. We will see in the next sections how can we select the variables α_i, in order to describe non–equilibrium Brownian motion of interacting particles,

2. MESOSCOPIC APPROACH TO THE DYNAMICS OF INTERACTING BROWNIAN PARTICLES

In this section we apply the scheme outlined above for the particular of Brownian motion of interacting Brownian particles. From the point of view of the formalism, we need to choose the relevant fluctuations involved in the problem. For Brownian particles, it is well known that the time scales involved with them, are well separated from the corresponding molecular time scales of the constituents of the supporting fluid (solvent). This physical fact is valid for a great variety of particles whose density (defined in this case as the ratio of the Brownian mass particle and the volume occupied by the particle) is greater than the density of the solvent (defined as the ratio of the total mass of the solvent and the total volume occupied by it). When this last condition is obeyed, we can choose with confidence the velocities $\vec{u}_i$ of the Brownian variables as the fluctuations $\vec{a}_i$ discussed above.

After the choice of the relevant fluctuating variables, we should take into account the direct interactions between the particles at least as additive by pairs. In this manner, the continuity eq. (4) is modified as

$$\frac{\partial P}{\partial t} + \sum_{i=1}^{N} u_i \frac{\partial P}{\partial r_i} - m^{-1} \sum_{i,j=1}^{N} \frac{\partial \phi_{ij}}{\partial r_i} \frac{\partial P}{\partial u_i} = - \sum_{i=1}^{N} \frac{\partial}{\partial u_i} J_{u_i}, \qquad (10)$$

where $-m^{-1}\frac{\partial \phi_{ij}}{\partial r_i}$ is the force over the particle i due to the direct interaction with the particle j and ϕ_{ij} is the intermolecular potential between the particles i and j. Following the method described above we can determine the kinetic description through the flux J_{u_i} (eq. 9),

$$J_{u_i} = - \sum_{\substack{j \neq i \\ j=1}}^{N} \beta_{ij} \left(P u_i + \frac{kT_{eq}}{m} \frac{\partial P}{\partial u_i} \right). \qquad (11)$$

and afterwards of its substitution in the continuity eq. 10, we obtain the corresponding N-particle Fokker–Planck equation. On the other hand, even if we can use the N–particle Fokker–Planck equation to construct the hydrodynamic equations for the set of Brownian particles, in order to have explicit relations for the thermodynamic and transport properties in terms of the direct interaction potential between the particles, we perform a contraction of the description at the two–particle level, i.e., after a systematic integration of the N–particle Fokker–Planck equation, we obtain the corresponding Fokker–Planck equations for the one and two-particle distribution functions [3], namely,

$$\frac{\partial}{\partial t} P^{(1)} + u_1 \cdot \frac{\partial}{\partial r_1} P^{(1)} - m^{-1} \int \frac{\partial \phi_{12}}{\partial r_1} \cdot \frac{\partial P^{(2)}}{\partial u_1} du_2 dr_2$$

$$= \frac{\partial}{\partial u_1} \cdot \beta_{11} \left(P_1^{(1)} u_1 + \frac{kT_{eq}}{m} \frac{\partial P^{(1)}}{\partial u_1} \right) \qquad (12)$$

and

$$\frac{\partial}{\partial t}P^{(2)} + \sum_{i=1}^{2} u_i \cdot \frac{\partial P^{(2)}}{\partial r_i} - m^{-1} \sum_{i,j=1}^{2} \frac{\partial \phi_{ij}}{\partial r_i} \cdot \frac{\partial P^{(2)}}{\partial u_i}$$

$$= \sum_{i,j=1}^{2} \frac{\partial}{\partial u_i} \cdot \beta_{ij} \left(P^{(2)}u_i + \frac{kT_{eq}}{m}\frac{\partial P^{(2)}}{\partial u_i} \right). \qquad (13)$$

where $P^{(1)}$ and $P^{(2)}$ are the one and two-particle distribution functions respectively, β_{11} stands for the usual friction coefficient (the same that was calculated by Stokes) for one particle and β_{ij} are the components of the two–particle friction coefficient that was first calculated by Batchelor [4].

With the above kinetic equations (12) and (13) we can construct the conservation equations for the Brownian particles, i.e., in order to obtain the momentum conservation, we multiply by the velocity u_1 the eq. 12 and afterwards of integrate over all the values of the velocities, we obtain the next balance equation,

$$\rho_B \frac{dv_{B_i}}{dt} = -\frac{\partial}{\partial r_i}\mathcal{P}_{B_{ij}} - \beta_{ii}\rho_B v_{B_i}. \qquad (14)$$

where ρ_B is the mass density and v_B the mean velocity of the Brownian particles, $\mathcal{P}_{B_{ij}} = \mathcal{P}_{B_{ij}}^k + \mathcal{P}_{B_{ij}}^\phi$ are components of the momentum pressure tensor with $\mathcal{P}_{B_{ij}}^k = m \int P^{(1)} (u_i - v_{B_i})(u_j - v_{B_j}) du$ the kinetic (non-interacting) part and

$$\mathcal{P}_{B_{ij}}^\phi = -\frac{m}{2} \int r_{ij} r_{ij} \frac{\phi_{ij}'(r_{ij})}{r_{ij}}$$

$$\times \left(\int_0^1 P^{(2)} (r_i - [1 - \alpha]r_{ij}, r_i + \alpha r_{ij}, u_i, u_j, t)\, d\alpha \right) dr_{ij} du_i du_j$$

the potential (interacting) part with $\phi_{ij}'(r_{ij}) = \frac{\partial \phi_{ij}}{\partial r_{ij}}$. This expressions for the pressure tensor are identical to the ones obtained earlier for the case of dense liquids [5], and similar to that case, from the decomposition of the pressure tensor $\mathcal{P}_{B_{ij}} = P_B \mathcal{I}_{ij} + \pi_{ij}$, where P_B is the osmotic pressure that the Brownian particles exerts over the solvent and π_{ij} is the viscous tensor, we obtain that $P_B = \frac{\rho_B}{m}kT - \frac{2\pi}{3}\left(\frac{\rho_B}{m}\right)^2 \int \frac{\partial \phi_{ij}}{\partial r} g_{eq}(r_{ij}) r_{ij}^3 dr_{ij}$, i.e. the osmotic pressure has the usual virial form where we need the information of the pair correlation function $g_{eq}(r_{ij})$ of the Brownian particles. The last term in the rhs of eq. (14) accounts for the momentum interchange of the particles with the solvent. One of the important relations arise when we account for the long time scales, i.e., $t >> \beta_{ii}^{-1}$. In this case we obtain from the momentum equation 14 the usual Fick's law $\rho_B v_{B_i} = -D_c \frac{\partial \rho_B}{\partial r_i}$ where D_c is the collective diffusion coefficient,

$$D_c = \frac{kT}{m} \left[1 - \frac{4}{3}\pi \frac{\rho_B}{kTm} \int_0^\infty \frac{\partial \phi_{12}}{\partial r} g_{eq}^{(2)} r_{12}^3 dr_{12} \right] \beta_{ii}^{-1}. \tag{15}$$

When we use the virial expression of osmotic pressure, the above last expression can be rewritten in a more compact form, namely,

$$D_c = \beta_{ii}^{-1} \left(\frac{\partial P_B}{\partial \rho_B} \right)_T, \tag{16}$$

i.e., we obtain from the momentum balance equation a generalized Einstein relation for interacting Brownian particles which involves to the collective diffusion coefficient D_c in terms of the mobility β_{ii}^{-1} (which is obtained from a hydrodynamical theory or a sedimentation experiment) and the osmotic compressibility $\left(\frac{\partial P_B}{\partial \rho_B} \right)_T$ (which can be obtained from a light scattering experiment).

To finish this section, we would like to emphasize the necessity to incorporate the potential conservative interactions between the particles in order to describe the energy balance equation properly. The internal energy $\rho_B u_B = \rho_B u_B^k + \rho_B u_B^\phi$, accounts for the kinetic (ideal) $\rho_B u_B^k = \frac{m}{2} \int P^{(1)} (u_i - v_B)^2 du_i$ and the potential (interacting) contributions $\rho_B u_B^\phi = \frac{m}{2} \int P^{(2)} \phi_{ij} du_i du_j dr_j$. In order to obtain the internal energy balance equation, we apply the Fokker–Planck eqs. (12) and (13), and the final result is

$$\frac{\partial}{\partial t} \rho_B u_B + \frac{\partial}{\partial r_i} \left(J_{q_i}^B + \rho_B u_B v_{B_i} \right) = -\frac{\rho_B}{m} v_{B_i} F_i + \mathcal{P}_{B_{ij}}$$

$$\frac{\partial v_{B_i}}{\partial r_i} = 2\rho_B \beta_{ii} \left(u_B^k - u_{B_{eq}}^k \right). \tag{17}$$

In this latter expression we should note that $-\frac{\rho_B(r_i,t)}{m} v_{B_i}(r_i,t) F_i(r_i,t) = \frac{\rho_B(r_i,t)}{m} v_{B_i} \int \frac{\rho_B(r_j,t)}{m} g^{(2)}(r_i,r_j,t) \frac{\partial \phi_{ij}}{\partial r_i} dr_j$ accounts for the rate at which dissipative work is done by a Brownian particle i over the solvent due to the pair effective long range interactions between two Brownian particles i and j. This term would be useful to describe for example the dissipative interaction between two Brownian macroions in solution. It is important to observe that in order to calculate this contribution, we need to know the dynamic pair correlation function $g^{(2)}(r_i, r_j, t)$ and the dynamic density profile around the two tracer particles, i.e., $\frac{\rho_B(r_i,t)}{m}$ and

$\frac{\rho_B(r_j,t)}{m}$. The second term in the rhs of eq. (17) accounts for the usual contribution to increase the internal energy due to the viscous shear rate and the last term is due to the interchange of energy between the particles and the solvent.

3. BROWNIAN MOTION UNDER A TEMPERATURE GRADIENT

One of the important issues in statistical mechanics concerns with the description of Brownian motion in a non–equilibrium heat bath. A system with an external fixed temperature gradient is a particular case of a non–equilibrium situation which consist of a stationary (time independent) state. For this case, the mesoscopic thermodynamic formalism permit to construct an N–particle Fokker–Planck equation [6] in an easier manner in comparison with the projector operator method [7]. Similarly to the case of the before section, a contraction in the description permit us to obtain the two–particle Fokker–Planck equation, namely,

$$\frac{\partial}{\partial t}P^{(2)} + \sum_{i=1}^{2} u_i \cdot \frac{\partial P^{(2)}}{\partial r_i} - m^{-1}\sum_{i,j=1}^{2}\frac{\partial \phi_{ij}}{\partial r_i}\cdot\frac{\partial P^{(2)}}{\partial u_i}$$

$$= \sum_{i,j=1}^{2}\frac{\partial}{\partial u_i}\cdot\beta_{ij}\left(P^{(2)}u_i + \frac{kT_{eq}}{m}\frac{\partial P^{(2)}}{\partial u_i} + \gamma_{ij}\frac{P^{(2)}}{T_j}\frac{\partial T}{\partial r_j}\right). \quad (18)$$

The main difference in comparison with the kinetic eq. (13) consist of the term $\gamma_{ij}\frac{P^{(2)}}{T_j}\frac{\partial T}{\partial r_j}$, which shows how the temperature gradient acts like an external force, where γ_{ij} stands for the thermal acceleration coefficient for the Brownian particles.

On the other hand, with the latter kinetic equation 18, we can construct the momentum balance, i.e.,

$$\rho_B\frac{dv_{B_i}}{dt} = -\frac{\partial}{\partial r_j}P_{B_{ij}} - \beta_{ij}\rho_B v_{B_j} + \frac{\gamma_{ij}}{T}\frac{\partial T}{\partial r_j}, \quad (19)$$

and similarly to the case without temperature gradient, when we take the long time behavior we obtain the mass flux

$$\rho_B v_{B_i} = -D_{c_{ij}}\frac{\partial\rho_B}{\partial r_j} - \frac{D_{T_{ij}}}{T}\frac{\partial T}{\partial r_j}. \quad (20)$$

Here $D_{c_{ij}} = \beta_{ij}^{-1}\left(\frac{\partial P_B}{\partial\rho_B}\right)_T$ is the collective diffusion coefficient and

$$D_{T_{ij}} = \beta_{ij}^{-1}\gamma_{ij} + \beta_{ij}^{-1}\left[\frac{\rho_B kT}{m} - \frac{2\pi T}{3}\left(\frac{\rho_B}{m}\right)^2\int_0^\infty\frac{\partial\phi_{12}}{\partial r}\frac{\partial g_{eq}}{\partial T}r^3 dr\right] \quad (21)$$

stands for the thermal diffusion coefficient which account for the change of the pair correlation function respect to the temperature, and the non–interacting case reduces to the expression

$$D_{T_{ii}} = \beta_{ii}^{-1} \left(\gamma_{ii} + \frac{\rho_B k}{m} \right) \qquad (22)$$

which is the scalar version of the tensorial form (21). On the other hand, using the above expression (22) with the ideal gas equation of state $p_B = \rho_B kT/m$, and if additionally we take into account that in the dilute non-interacting case $D_c = D$, here D is the self diffusion coefficient, we obtain the next expression,

$$D_T = \rho_B D \left(1 + \frac{\gamma m}{kT} \right), \qquad (23)$$

i.e., a relationship reported previously [8].

To finish this report, we would like to remark that the thermodynamic mesoscopic approach is not restricted to the study of transport properties of colloidal-like systems. The important feature is to identify the relevant fluctuations in the system, which are events that take place on a time scale much slower in comparison to molecular or microscopic times, and work with them as internal thermodynamic variables in order to obtain their corresponding Fokker–Planck equations.

Acknowledgments

The author wish to express his gratitude to CONACyT for the financial support through the grant J32094–E.

References

[1] P. Mazur, *Physica A* **274** (1999) 491.

[2] S.R. de Groot and P. Mazur, *Nonequilibrium Thermodynamics*, (Dover publications, 1984).

[3] M. Mayorga, L. Romero-Salazar and J. M. Rubí, *Physica A* (in press).

[4] G.K. Batchelor, *J. Fluid Mech.* **52** (1972) 245.

[5] J.H.Irving and J.G. Kirkwood, *J. Chem. Phys.* **18** (1959) 817.

[6] J.M. Rubí and P. Mazur, *Physica A* **250** (1998) 253.

[7] J. M. Shea, and I. Oppenheim, *J. Phys. Chem.* **100** *(1996)* 19035. idem *Physica A* **247** (1997) 417.

[8] A. Pérez-Madrid, J.M. Rubí and P. Mazur, *Physica A* **212** (1994) 231.

CHAOS AND TRANSPORT IN DETERMINISTIC INERTIAL RATCHETS

José L. Mateos

Instituto de Física,
Universidad Nacional Autónoma de México,
Apartado Postal 20-364, 01000 México, D.F., México
E-mail: mateos@fisica.unam.mx

Abstract   We address the problem of the chaotic transport of particles in an asymmetric periodic ratchet potential. We briefly discuss the transport of Brownian particles far from equilibrium and consider in detail the limiting case of vanishing thermal noise. In this deterministic limit, the inertial term in the equation of motion plays a crucial role, since the dynamics can become chaotic and modify the transport properties. We discuss the dynamical origin of current reversals in ratchets and possible applications.

Keywords: Chaotic transport; ratchets; Brownian motors

1.   INTRODUCTION

Recent advances in non–equilibrium statistical physics have revealed various instances of the surprising phenomenon of noise enhanced order, such as stochastic resonance [1], Brownian motors or noise–induced transport [2]. These remarkable phenomena occur due to the constructive role of noise in nonlinear dynamical systems.

Noise–induced, directed transport in a spatially periodic system in thermal equilibrium is ruled out by the second law of thermodynamics. Therefore, in order to generate transport, the system has to be driven away from thermal equilibrium by an additional deterministic or stochastic perturbation. In the most interesting situation, these perturbations are unbiased, that is, the temporal, spatial or ensemble averages of the associated forces are required to vanish. Besides the breaking of thermal equilibrium, another important requirement to get directed transport in a spatially periodic system is the breaking of the spatial inversion symmetry. We speak then of Brownian motors, ratchet devices, or, in the

Developments in Mathematical and Experimental Physics,
Volume B: Statistical Physics and Beyond
Edited by Macias *et al.*, Kluwer Academic/Plenum Publishers, 2003

71

biological realm, of molecular motors. These so–called ratchets can be modelled, for instance, by considering the nonlinear dynamics of a particle in a spatially periodic asymmetric potential and acted upon by an external time–dependent force of zero average.

The recent burst of interest in this area is motivated by a series of scientific and possible technological applications in devising and understanding noise–induced separation and pumping techniques. The activity has been so intense that in less than a decade several hundred papers in this area have been written. See the review article by P. Reimann [3] for a thorough review of the field.

2.   DETERMINISTIC RATCHETS

Although the vast majority of the literature in this field considers the presence of noise, very recently there has been some interest in understanding in detail the transport properties of classical *deterministic inertial ratchets* [4, 5, 6, 7, 8, 9, 10, 11, 12, 13, 14, 15, 16, 17, 18, 19]. These ratchets have in general a chaotic dynamics that modifies the transport properties. The implications of this chaotic dynamics in deterministic ratchets have recently been extended to the quantum domain [20]. Also, some authors have consider the Hamiltonian limit of vanishing dissipation, together with the possible connection with quantum chaos [21].

In particular, in a recent paper [5], we established a connection between the current generated in a deterministic rocking ratchet and the bifurcation diagram associated with the chaotic dynamics. When we plot the current, defined as an average velocity, against a control parameter of the system, we found multiple current reversals; on the other hand, we obtained the bifurcation diagram for the velocity as a function of the same control parameter. We noticed a clear connection between these two plots and by doing a close comparison we found that the current reversals occur associated with bifurcations. In many cases, these reversals take place at crisis bifurcations from a chaotic to a periodic regime.

It is worth mentioning that this prediction has been verified qualitatively in a recent experiment done by Carapella *et al.* [16], using the ratchet effect for a relativistic flux quantum trapped in an annular Josephson junction embedded in an inhomogeneous magnetic field.

In another paper [7], we elaborate on this idea by studying current reversals and bifurcations for point particles and extended rigid dimers. We analyzed trajectories and orbits in phase space for point particles and we showed current reversals, as a function of the size of the rigid dimers, together with the corresponding bifurcation diagram.

3. THE RATCHET POTENTIAL MODEL

Let us consider the same equation of motion studied in [5, 6, 7], that is, a one–dimensional problem of a particle driven by a periodic time–dependent external force, under the influence of an asymmetric periodic potential of the ratchet type. The time average of the external deterministic force is zero. Here, we do not take into account any kind of noise, and thus the dynamics is deterministic. The equation of motion is given by

$$m\ddot{x} + \gamma\dot{x} + \frac{dV(x)}{dx} = F_0 \cos(\omega_D t), \tag{1}$$

where m is the mass of the particle, γ is the friction coefficient, $V(x)$ is the external asymmetric periodic potential, F_0 is the amplitude of the external force and ω_D is the frequency of the external driving force. The ratchet potential is given by

$$V(x) = V_1 - V_0 \sin \frac{2\pi(x - x_0)}{L} - \frac{V_0}{4} \sin \frac{4\pi(x - x_0)}{L}, \tag{2}$$

where L is the periodicity of the potential, V_0 is the amplitude, and V_1 is an arbitrary constant. The potential is shifted by an amount x_0 in order that the minimum of the potential is located at the origin.

Let us define the following dimensionless units: $x' = x/L$, $x'_0 = x_0/L$, $t' = \omega_0 t$, $w = \omega_D/\omega_0$, $b = \gamma/m\omega_0$ and $a = F_0/mL\omega_0^2$. Here, the frequency ω_0 is given by $\omega_0^2 = 4\pi^2 V_0 \delta/mL^2$ and δ is defined by $\delta = \sin(2\pi|x'_0|) + \sin(4\pi|x'_0|)$.

The frequency ω_0 is the frequency of the linearized motion around the minima of the potential, thus we are scaling the time with the natural period of motion $\tau_0 = 2\pi/\omega_0$. The dimensionless equation of motion, after renaming the variables again without the primes, becomes

$$\ddot{x} + b\dot{x} + \frac{dV(x)}{dx} = a \cos(wt), \tag{3}$$

where the dimensionless potential can be written as

$$V(x) = C - \frac{1}{4\pi^2\delta} \left[\sin 2\pi(x - x_0) + \frac{1}{4} \sin 4\pi(x - x_0) \right] \tag{4}$$

and is depicted in Fig. 1. The constant C is such that $V(0) = 0$, and is given by $C = -(\sin 2\pi x_0 + 0.25 \sin 4\pi x_0)/4\pi^2\delta$. In this case, $x_0 \simeq -0.19$, $\delta \simeq 1.6$ and $C \simeq 0.0173$.

In the equation of motion Eq. (3) there are three dimensionless parameters: a, b and w, defined above in terms of physical quantities. The parameter $a = F_0/mL\omega_0^2$ is the ratio of the amplitude of the external

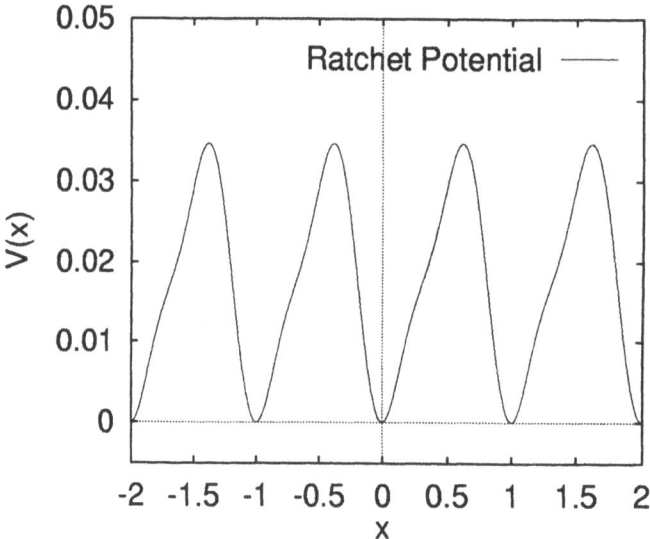

Figure 1.   The dimensionless ratchet periodic potential $V(x)$.

force and the force due to the potential $V(x)$. This can be seen more clearly using the expression for ω_0^2 in terms of the parameters of the potential. In this case, the ratio becomes

$$a = \frac{1}{4\pi^2 \delta} \frac{F_0 L}{V_0},\tag{5}$$

that is, except for a constant factor, a is the ratio of F_0 and the force V_0/L, where V_0 is the amplitude and L the periodicity of the potential (see Eq. (2)). The parameter b is simply the dimensionless friction coefficient, and w is the ratio of the driving frequency of the external force and ω_0.

The extended phase space in which the dynamics is taking place is three–dimensional, since we are dealing with an inhomogeneous differential equation with an explicit time dependence. This equation can be written as a three–dimensional dynamical system, that we solve numerically, using the fourth–order Runge–Kutta algorithm. The equation of motion Eq. (3) is nonlinear and thus allows the possibility of periodic and chaotic orbits. If the inertial term associated with the second derivative $\ddot{x}$ were absent, then the dynamical system could not be chaotic.

The main motivation behind this work is to study in detail the origin of the current reversal in a chaotically deterministic rocked ratchet as found in [5]. In order to do so, we have to study first the current J itself, that we define as the time average of the average velocity over an ensemble of initial conditions. Therefore, the current involves two different averages: the first average is over M initial conditions, that we take equally distributed in space, centered around the origin and with an initial velocity equal to zero. This quantity is a single number for a fixed set of parameters a, b, w.

Besides the orbits in the extended phase space, we can obtain the Poincaré section, using as a stroboscopic time the period of oscillation of the external force. With the aid of Poincaré sections we can distinguish between periodic and chaotic orbits, and we can obtain a bifurcation diagram as a function of the parameter a. As was shown in [5], there is a connection between the bifurcation diagram and the current.

4.   NUMERICAL RESULTS

Using the definition of the current J given in the previous section, we calculate J fixing the parameters $b = 0.1$ and $w = 0.67$ and varying the parameter a. The current shows, as stressed before [4, 5, 6, 7], multiple current reversals and a complex variation with a, as shown in Fig. 2. We can observe strong fluctuations as well as portions where the current is approximately constant. The challenge here is to explain this high variability in the current with the aid of what we know from the nonlinear chaotic dynamics of the system.

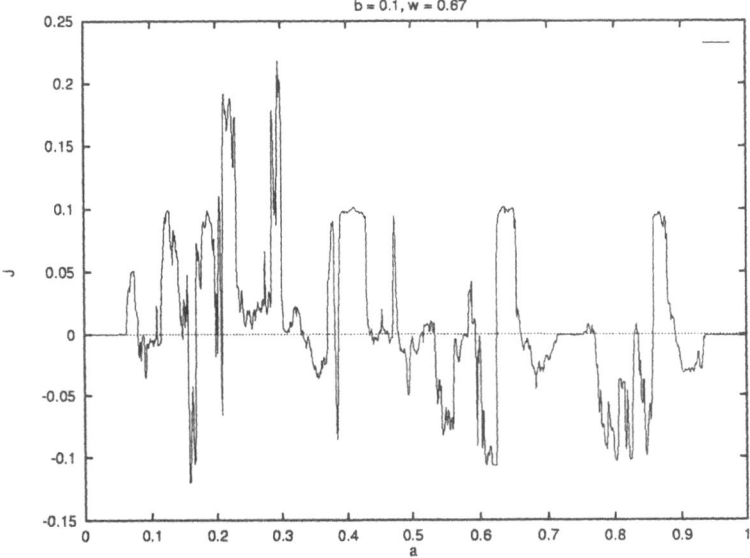

Figure 2.   For $b = 0.1$ and $w = 0.67$ we show the current J as a function of a. We can see multiple current reversals.

In order to understand the first reversal of the current, let us analyze a range of values of a between 0.065 and 0.085; we chose $b = 0.1$ and $w = 0.67$. In Fig. 3a we show the bifurcation diagram as a function of a, and in Fig. 3b we depict the current in the same range of a. Let us imagine that an ensemble of particles are initially located at the minimum of the ratchet potential around the origin, and that all these particles have an initial velocity equal to zero. For $a = 0$, we have no external force and thus, all these particles remain in the minimum around the origin and therefore the current is zero. For very small values of a, we still have a zero current, since the particles have friction and tend to oscillate in this minimum. However, there is a critical value of a for which the particles start to overcome the potential barriers around the minimum and transport along the ratchet potential in a periodic or chaotic way. At the beginning, the current is dominated by transport due to periodic orbits, but for larger values of a, some of the orbits in the ensemble become chaotic and the transport is not as efficient as before, resulting in a current that starts to oscillate erratically. In fact, in this region, there exist the possibility of coexistence of multiple attractors in the phase space. For example, in Fig. 3a, we have two coexistent attractors: a periodic and a chaotic one, around $a = 0.067$. In this case, depending on the initial conditions, some orbits in the ensemble can end up in the periodic attractor, and the rest in the chaotic attractor. In general, the current can depend on which basin of attraction one is located, because different attractors have different transport properties. For example, one attractor can transport particles in the positive direction, while the other attractor transport particles in the negative direction. The case of coexisting attractors is analyzed in [8]. For values of a even larger, all the orbits enter a chaotic region through a period–doubling bifurcation, and the current starts to decrease inside this chaotic band. Finally, exactly at the bifurcation point where a periodic window opens, the current drops to zero and becomes negative in a very abrupt way. Let us focus first on the range of the control parameter where the first current reversal takes place. This occurs around $a \simeq 0.08$ as shown in Fig. 3. We can observe a period–doubling route to chaos and after a chaotic region, there is a crisis bifurcation (see [22]) taking place at the critical value $a_c \simeq 0.08092844$. It is precisely at this bifurcation point that the current reversal occurs. After this bifurcation, a periodic window emerges, with an orbit of period four. In Figs. 3a,b we are analyzing only a short range of values of a, where the first current reversal takes place. If we vary a further, we can obtain multiple current reversals, as shown in Fig. 2.

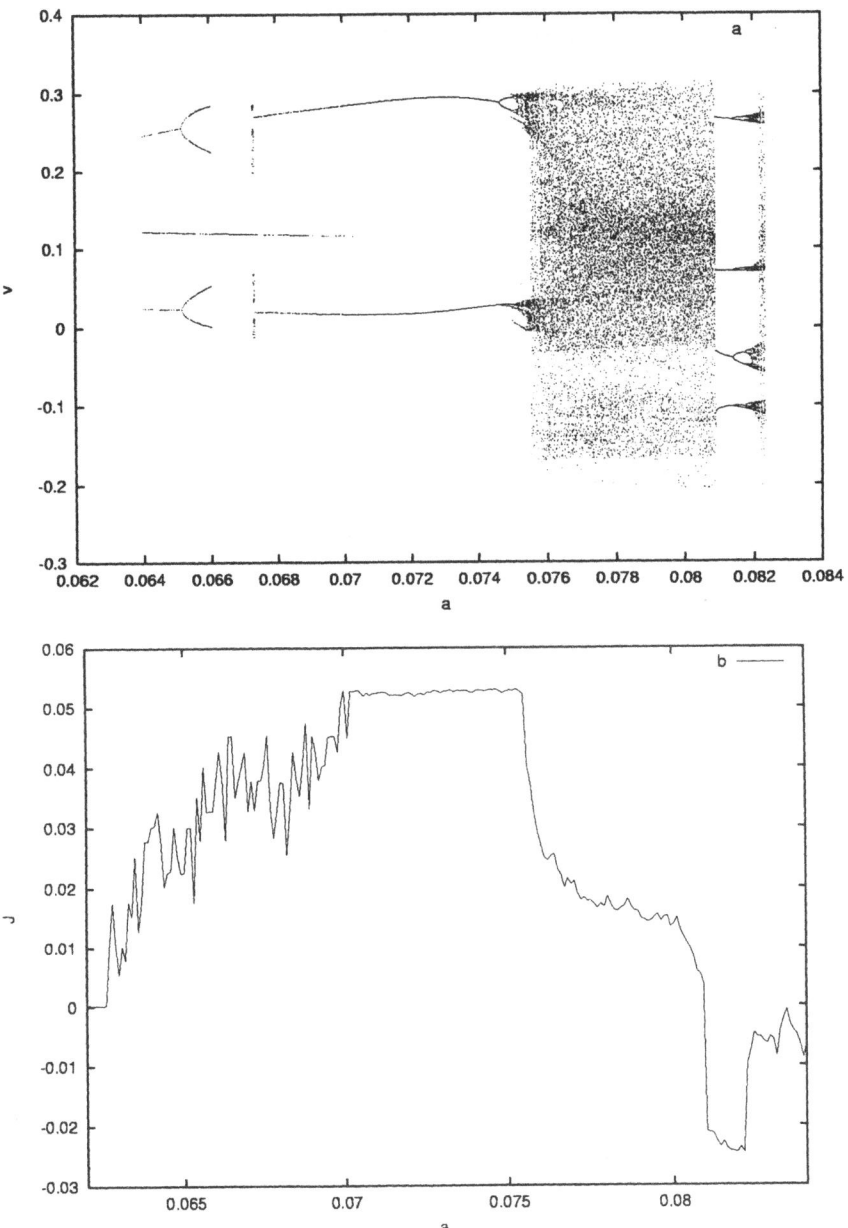

Figure 3. For $b = 0.1$ and $w = 0.67$ we show: (a) The bifurcation diagram as a function of a, and (b) The current J as a function of a. The range in the parameter a corresponds to the first current reversal.

In Fig. 4a, we show a typical trajectory for a just below a_c. The trajectory is chaotic and the corresponding chaotic attractor is depicted

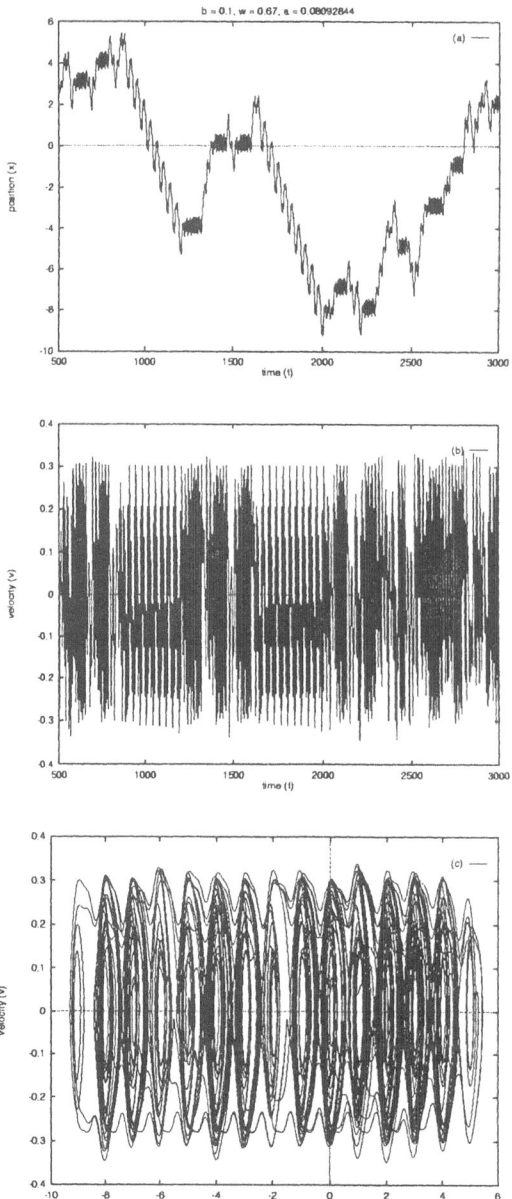

Figure 4.   For $b = 0.1$ and $w = 0.67$ and $a = 0.08092844$ we show: (a) The trajectory of the particle as a function of time, (b) the velocity as a function of time and (c) the phase space. This case corresponds to a near the bifurcation, where the dynamics becomes intermittent and there is anomalous diffusion.

in Fig. 5. In this case, the particle starts at the origin with no velocity; it jumps from one well in the ratchet potential to another well to the right or to the left in a chaotic way. The particle gets trapped, oscillating for a while in a minimum (sticking mode), as is indicated by the integer values of x in the ordinate, and suddenly starts a running mode with average constant velocity in the negative direction. In terms of the velocity, these running modes, correspond to periodic motion. This can be seen more clearly in Fig. 4b, where we plot the velocity as a function of time in the same range of values as the orbit in Fig. 4a. In Fig. 4c we show the corresponding phase space.

The phenomenology can be described as follows. For values of a above a_c, the attractor is a periodic orbit. For a slightly less than a_c there are long stretches of time (running or laminar modes) during which the orbit appears to be periodic and closely resembles the orbit for $a > a_c$, but this regular (approximately periodic) behavior is *intermittently* interrupted by finite duration "bursts" in which the orbit behaves in a chaotic manner. The net result in the velocity is a set of periodic stretches of time interrupted by burst of chaotic motion, signaling precisely the phenomenon of *intermittency* [22]. As a approach a_c from below, the duration of the running modes in the negative direction increases, until the duration diverges at $a = a_c$, where the trajectory becomes truly periodic.

To complete this picture, in Fig. 5, we show two attractors: (1) the chaotic attractor for $a = 0.08092$, just below a_c, corresponding to the trajectory in Fig. 4a, and (2) the period-4 attractor for $a = 0.08093$, corresponding to the period-4 window in the bifurcation diagram. This periodic attractor consists of four points in phase space, which are located at the center of the open circles. We obtain these attractors confining the dynamics in x between -0.5 and 0.5, that is, we used the periodicity of the potential $V(x + 1) = V(x)$, to map the points in the x axis modulo 1. Thus, even though the trajectory transport particles to infinity, when we confine the dynamics, the chaotic structure of the attractor is apparent. As a approaches a_c from below, the dynamics in the attractor becomes intermittent, spending most of the time in the vicinity of the period–4 attractor, and suddenly "jumping" in a chaotic way for some time, and then returning close to the period–four attractor again, and so on. In terms of the velocity, the result is an intermittent time series as the one depicted in Fig. 4b. In order to characterize the deterministic diffusion in this regime, we calculate the mean square displacement $\langle (x - \langle x \rangle)^2 \rangle$ as a function of time. We obtain numerically that $\langle (x - \langle x \rangle)^2 \rangle \sim t^\alpha$, where the exponent $\alpha \simeq 3/2$. This is a signature of anomalous deterministic diffusion, in which $\langle (x - \langle x \rangle)^2 \rangle$ grows faster than linear, that is, $\alpha > 1$ (superdiffusion). Normal deterministic diffusion corresponds to $\alpha = 1$. In contrast, the trajectories in the periodic windows transport particles in a ballistic way, with $\alpha = 2$. The relationship

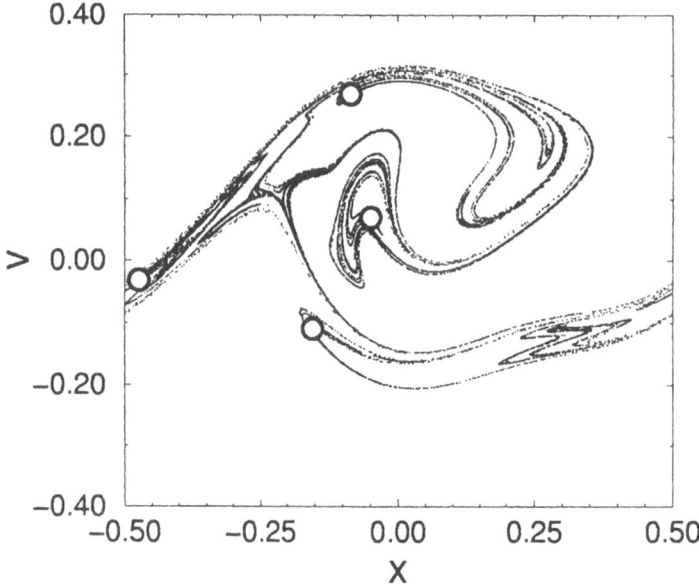

Figure 5. For $b = 0.1$ and $w = 0.67$ we show two attractors: a chaotic attractor for $a = 0.08092$, just below a_c, and a period–four attractor, for $a = 0.08093$, consisting of four points located at the center of the open circles.

between anomalous deterministic diffusion and intermittent chaos has been explored recently, together with the connection with Lévy flights. See the reviews [23, 24, 25]. The signature of this anomalous diffusion can be seen more clearly in Fig. 4a, where we plot a trajectory showing running modes of different sizes, typical of Lévy flights.

5.   EXTENDED OBJECTS: RIGID DIMERS

Let us consider now the case where we have, instead of a point particle, an extended rigid dimer of length d, that is, two point particles kept rigidly at a fixed distance d. This case was studied in [7]. If we define a dimensionless length $l = d/L$, where L is the periodicity of the ratchet potential in Eq. (2), the dynamics of this dimer in the ratchet potential is equivalent to the dynamics of a point particle in an effective potential given by

$$U_l(x) = \frac{1}{2}\left[V(x) + V(x + l)\right] \tag{6}$$

where $V(x)$ is the dimensionless potential in Eq. (4). Due to the periodicity of the dimensionless ratchet potential $V(x + 1) = V(x)$, we have that the effective dimensionless potential is periodic: $U_l(x + 1) = U_l(x)$. Since the period is one, it is enough to take the dimensionless length of the rigid dimer in the unit interval: $0 \leq l \leq 1$. We can see that we recover the point particle case when $l = 0$ or $l = 1$. The equation of motion for this dimer is then given by

$$\ddot{x} + b\dot{x} + \frac{dU_l(x)}{dx} = a\cos(wt), \tag{7}$$

In this case, we have an additional parameter l in the dynamics. According to this equation of motion, the dynamics of the dimer is modelled using a single degree of freedom, since we are not considering any internal degree of freedom or rotational motion.

It is important to mention that several authors before had studied the case of a rigid dimer in a ratchet potential, in the overdamped limit. The case of a single rigid dimer in the overdamped limit in the presence of noise was studied in [26, 27]. They obtain a current reversal as a function of the dimer length. Also, in the overdamped limit, there are studies of the collective transport of N rigid rods of finite size; these rods are interacting through hard core repulsion, [28, 29]. For an overview of the collective transport in ratchets, see [3, 30]. More recently, the problem of horizontal transport of vertically vibrated granular layers on a ratchet was addressed [31, 32]; it was found that increasing the layer thickness leads to a reversal of the current. In [33] the authors studied the case of granular binary mixtures where, unlike other segregation phenomena, in which the segregation is due to the collective behavior of the grains, the interaction between the ratchet and the individual particles is dominant. If the density of particles is low, the interaction between the objects can be neglected, recovering a single rod description. Finally, in [34], the authors perform simulations of elongated grains on a vibrating ratchet-shaped base; they consider grains composed of one (monomers), two (dimers) or three (trimers) colinear spheres. For some parameters, they found that monomers and dimers move in opposite directions.

The underdamped determinist dynamics of a rigid dimer was analyzed in [7]. In that paper, we calculate the bifurcation diagram as a function of the size l in the unit interval, and the corresponding average velocity in the same range. We obtained a rich structure of bifurcations when we vary the size of the rigid dimer. We noticed the presence of periodic and chaotic orbits. As a consequence of the complexity of the bifurcation diagram, we obtain an average velocity with fine structure. Again, we obtained a clear connection between these two graphs.

In particular, we can found a situation were we have a positive current for one size l_1, and a negative current for another size l_2, even for a small difference between these two sizes. Thus, if we have a granular medium with a mixture of these two types of grains, then using a single ratchets we can separate the two species in a very efficient way. So, we can think of a possible mechanism for segregation of particles according to its size, even for the same mass.

6.   CONCLUDING REMARKS

In summary, we have briefly discussed the dynamics of Brownian motors, that is Brownian particles in presence of an external ratchet potential. We emphasize the limiting case of vanishing noise in which we arrive at a deterministic dynamics. In this limit, we were able to understand the role that the chaotic dynamics plays in the transport of particles. In particular we established a connection between the bifurcation diagram and the current and identified the mechanism by which the current reversals in deterministic ratchets arise: it corresponds to bifurcations, usually from a chaotic to a periodic regime. Near this crisis bifurcations, the chaotic trajectories exhibit intermittent chaos and the transport arises through deterministic anomalous diffusion with an exponent greater than one. The connection between bifurcations and current reversals is a rather general connection: it occurs for several control parameters, like the amplitude and frequency of forcing, friction, the size of an extended dimer, among others. The richness of the bifurcation diagrams in this ratchet systems, urges us to explore this connection in more detail in the near future.

Acknowledgments

The author gratefully acknowledges financial support from UNAM through project DGAPA–IN–111000.

References

[1] L. Gammaitoni, P. Hänggi, P. Jung, and F. Marchesoni, Rev. Mod. Phys. **70** (1998) 223; R. D. Astumian and F. Moss, Chaos **8** (1998) 533.

[2] P. Hänggi and R. Bartussek, in "Nonlinear Physics of Complex Systems", Lecture Notes in Physics, Vol. **476**, edited by J. Parisi, S. C. Müller, and W. Zimmermann. (Springer, Berlin, 1996) pp. 294; R. D. Astumian, Science **276** (1997) 917; F. Jülicher, A. Ajdari, and J. Prost, Rev. Mod. Phys. **69** (1997) 1269;

[3] P. Reimann, Phys. Rep. **361** (2002) 57.

[4] P. Jung, J. G. Kissner, and P. Hänggi, Phys. Rev. Lett. **76** (1996) 3436.

[5] J. L. Mateos, Phys. Rev. Lett. **84** (2000) 258.

[6] J. L. Mateos, Acta Phys. Polonica **32** (2001) 307.

[7] J. L. Mateos, Physica D (2002), to be published.

[8] J. L. Mateos, Phys. Rev. E (2002), submitted.

[9] M. Porto, M. Urbakh, and J. Klafter, Phys. Rev. Lett. **84** (2000) 6058; *ibid.*, **85** (2000) 491; M. Porto, Phys. Rev. **E63** (2001) 030102.

[10] S. Flach, O. Yevtushenko, and , Y. Zolotaryuk, Phys. Rev. Lett. **84** (2000) 2358; O. Yevtushenko, S. Flach, and K. Richter, Phys. Rev. **E61** (2000) 7215; O. Yevtushenko, S. Flach, Y. Zolotaryuk, and A. A. Ovchinnikov, Europhys. Lett. **54** (2001) 141; S. Denisov and S. Flach, Phys. Rev. **E64** (2001) 056236.

[11] M. Barbi and M. Salerno, Phys. Rev. **E62** (2000) 1988; *ibid.*, **E63** (2001) 066212.

[12] C. M. Arizmendi, Fereydoon Family, and A. L. Salas–Brito, Phys. Rev. **E63** (2000) 061104; H. A. Larrondo, Fereydoon Family, and C. M. Arizmendi, Physica **A303** (2002) 67.

[13] B. Tilch, F. Schweitzer, and W. Ebeling, Physica **A273** (1999) 294; F. Schweitzer, B. Tilch, and W. Ebeling, Eur. Phys. J. **B14** (2000) 157.

[14] S. Cilla, F. Falo, and L. M. Floría, Phys. Rev. **E63** (2001) 031110.

[15] G. Carapella, Phys. Rev. **B63** (2001) 054515; G. Carapella and G. Costabile, Phys. Rev. Lett. **87** (2001) 077002.

[16] G. Carapella, G. Costabile, R. Latempa, N. Martucciello, M. Cirillo, A. Polcari, and G. Filatrella, cond-mat/0112467 (27 Dec 2001). To appear in Physica C.

[17] E. Goldobin, A. Sterck, and D. Koelle, Phys. Rev. **E63** (2001) 031111.

[18] B. Nordén, Y. Zolotaryuk, P. L. Christiansen, A. V. Zolotaryuk, Phys. Rev. **E65**, (2001) 011110.

[19] M. Borromeo, G. Costantini, and F. Marchesoni, Phys. Rev. **E65** (2002) 041110.

[20] P. Reimann, M. Grifoni and P. Hänggi, Phys. Rev. Lett. **79** (1997) 10.

[21] T. Dittrich, R. Ketzmerick, M. -F. Otto, and H. Schanz, Ann. Phys. (Leipzig) **9** (2000) 755; H. Schanz, M.-F. Otto, R. Ketzmerick, and T. Dittrich, Phys. Rev. Lett. **87** (2001) 070601; I. Goychuk and P. Hänggi, J. Phys. Chem. **B105**, (2001) 6642; H. Linke, T. E. Humphrey, R. P. Taylor, A. P. Micolich, and R. Newbury, Physica Scripta **T90** (2001) 54.

[22] E. Ott, Chaos in Dynamical Systems. (Cambridge University Press, 1993).

[23] M. F. Shlesinger, G. M. Zaslavsky, J. Klafter, Nature (London) **363** (1993) 31.

[24] M. F. Shlesinger, G. M. Zaslavsky, J. Klafter, Physics Today **49** (1996) 33.

[25] T. Geisel, in "Lévy Flights and Related Topics in Physics", Lecture Notes in Physics, Vol. **450**, edited by M. F. Shlesinger, G. M. Zaslavsky, U. Frisch (Springer, Berlin, 1995) pp. 153.

[26] T. E. Dialynas, G. P. Tsironis, Phys. Lett. A **218** (1996) 292.

[27] T. E. Dialynas, Katya Lindenberg, G. P. Tsironis, Phys. Rev. **E56** (1997) 3976.

[28] I. Derényi, T. Vicsek, Phys. Rev. Lett. **75** (1995) 374.

[29] I. Derényi, A. Ajdari, Phys. Rev. **E54** (1996) R5.

[30] I. Derényi, P. Tegzes, T. Vicsek, Chaos 8(1998) 657.

[31] Z. Farkas, P. Tegzes, A. Vukics, T. Vicsek, Phys. Rev. **E60** (1999) 7022.

[32] M. Levanon, D. C. Rapaport, Phys. Rev. **E64** (2001) 011304.

[33] Z. Farkas, F. Szalai, D. E. Wolf, T. Vicsek Phys. Rev. **E65** (2002) 022301.

[34] J. F. Wambaugh, C. Reichhardt, C. J. Olson, Phys. Rev. **E65** (2002) 031308.

II

BIOLOGICAL SYSTEMS AND POLYMER

COMPUTER SIMULATION MEETS MOLECULAR BIOLOGY

Volkhard Helms, Christian Gorba, Markus Lill
Max Planck Institute of Biophysics, Kennedyallee 70,
60596 Frankfurt, Germany
vhelms@mpibp-frankfurt.mpg.de

Abstract   Computer simulation is a quickly expanding discipline that is entering fields as diverse as astrophysics, polymer physics, chemical physics, and even molecular biology and biochemistry. Our own field is theoretical biophysics where we collaborate with a number of experimental biological groups. Although standard algorithms have emerged since the 1960s, and have been implemented in a number of freely available or commercial simulation packages, there is a constant need for new developments that open up new areas of research. Here, two approaches are described where standard simulation methodology is being expanded to treat diffusional dynamics of multi-protein systems and to treat proton transport in macromolecules.

Keywords: Molecular dynamics, Brownian dynamics, photosynthesis, proton transport.

1.   INTRODUCTION

Photosynthesis and respiration are fundamental bioenergetic processes in biological cells that are the basis for life on earth. The complex machinery includes photoexcitation, transfer of electrons and protons, and several chemical reactions. All these steps take place in large integral membrane proteins, the three–dimensional structures of which were characterized over the past 15 years by X–ray and electron diffraction. In the first part of this report we focus on the diffusion of cytochrome c, a water–soluble electron carrier protein, between the bacterial reaction centre and the cytochrome bc_1 complex. The process can be simplified in the following way: cytochrome c starts by picking up one electron at the bc_1 complex and diffuses to the reaction centre to deliver its electron. A real biological cell contains a large number of cytochrome c and membrane proteins which makes it a many-particle system. See [1] for the

Developments in Mathematical and Experimental Physics,
Volume B: Statistical Physics and Beyond
Edited by Macias *et al.*, Kluwer Academic/Plenum Publishers, 2003

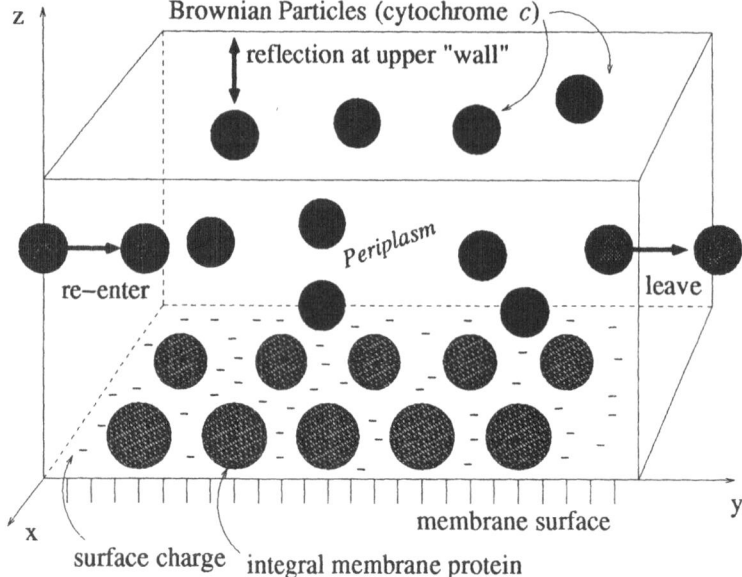

Figure 1. Model of the simulation box. The periodically distributed photosynthetic units are depicted as grey spheres on the membrane surface. The simulation uses periodic boundary conditions in x,y direction, whereas the diffusing particles (black) are forced not to leave the box in z direction. This is achieved by simply reflecting them at the upper wall.

first stochastic dynamics simulation of a comparable cellular system. By studying the diffusional aspects of the photosynthetic system, we want to answer question like: what is the concentration profile of cytochrome c above the oppositely charged membrane, or what is the rate of contacts between cytochrome c's and the membrane proteins, i.e., how often is an electron picked up and delivered to the reaction centre?

2. DIFFUSIONAL DYNAMICS OF CYTOCHROME *C* MOLECULES STUDIED BY BROWNIAN DYNAMICS SIMULATIONS

The diffusing cytochrome$c$ molecules are treated as spherical particles that move in the electrostatic field of the membrane with embedded, immobile, membrane proteins, see Figure 1.

The Brownian motion of the diffusing particles is best described by the Brownian dynamics simulation method that is based on the Langevin equation for a single particle. For N particles there are $6N$ equations of motion, $3N$ for translation and $3N$ for rotation, respectively. Together they form the set of **Langevin equations**:

$$m_i \dot{v}_i = -\sum_{j=1}^{3N} \left(\Gamma_{ij}^T \cdot v_j + \Gamma_{ij}^{TR} \cdot \omega_j \right) + \vec{F}_i + \sum_{j=1}^{6N} \alpha_{ij} \cdot f_j \qquad (1)$$

$$I_i \dot{\omega}_i = -\sum_{j=1}^{3N} \left(\Gamma_{ij}^{RT} \cdot v_j + \Gamma_{ij}^R \cdot \omega_j \right) + \vec{T}_i + \sum_{j=1}^{6N} \alpha_{(3N+i)j} \cdot f_j, \qquad (2)$$

with $v_i = \dot{r}_i$ (spatial coordinates) and $\omega_i = \dot{\phi}_i$ (rotation angle about i–th axis). These equations contain three types of interactions:

- **hydrodynamic interactions** : described by the friction matrices Γ that are related to the so–called diffusion matrices D by

$$D_{ij} = \beta^{-1} \left(\Gamma^{-1} \right)_{ij}, \; \beta = (k_B T)^{-1}, \qquad (3)$$

The matrices α_{ij} for the random forces (see below) are connected to the friction matrices by:

$$\Gamma_{ij} = \beta \sum_{l=1}^{6N} \alpha_{il} \cdot \alpha_{jl}, \qquad (4)$$

- **systematic forces** : external forces and interparticle forces F_i and T_i (e.g. Coulomb interaction)

- **random forces**: account for properly adjusted stochastic collisions with solvent molecules.

Starting from the Langevin equation Ermak and McCammon [2] derived an iteration algorithm for the position coordinates of the particles which is valid on the over–damped Brownian time scale for time steps $\Delta t \gg \beta m D$. A generalized algorithm for both translation and rotation was developed in [3].

$$y_i = y_i^0 + \sum_{j=1}^{6N} \left(\frac{\partial D_{ij}^0}{\partial y_j} \right) \Delta t + \beta \sum_{j=1}^{6N} D_{ij}^0 \cdot \mathcal{F}_j^0 \Delta t + \vec{R}_i(D_{ij}^0, \Delta t), \qquad (5)$$

where the subscript 0 means that the value has to be computed at the beginning of each time step. y and $\mathcal{F}$ are 6N–dimensional vectors of the generalized coordinates and of the generalized force. The 6N–dimensional random displacement vector R_i contains the displacements due to the random forces and has the properties

$$\langle R_i \rangle = 0, \; i \leq 6N \qquad (6)$$

$$\langle R_i(t) R_j(t') \rangle = 2 D_{ij}^0 \delta(t - t'). \tag{7}$$

The electrostatic interparticle forces are modelled as screened Coulombic interactions:

$$
\begin{aligned}
\vec{F}_{cms} &= \sum_{i=1}^{n} \vec{F}_{i,total} = \sum_{i=1}^{n} \sum_{j=1}^{m} \vec{F}_{ij} = \\
&\sum_{i=1}^{n} \sum_{j=1}^{m} \frac{q_i q_j}{4\pi\epsilon\epsilon_0} \left(\frac{\vec{r}_i - \vec{r}_j}{|\vec{r}_i - \vec{r}_j|^3} + \frac{\vec{r}_i - \vec{r}_j}{l_D |\vec{r}_i - \vec{r}_j|^2} \right) e^{-\frac{|\vec{r}_i - \vec{r}_j|}{l_D}} .
\end{aligned}
\tag{8}
$$

The torque with respect to the center of mass of a sphere is then

$$\vec{T}_{cms} = \sum_{i=1}^{n} (\vec{r}_i - \vec{r}_{cms}) \times \vec{F}_{i,total}. \tag{9}$$

2.1. HYDRODYNAMICS

The hydrodynamic interaction is described by the grand diffusion tensor, which consists of three parts: the translational, rotational and coupled rotational–translational diffusion tensors. A general scheme to evaluate diffusion tensors of an arbitrary number of spheres, immersed in a viscous fluid, is presented in [4].

$$D = \begin{pmatrix} D^T & D^{TR} \\ D^{RT} & D^R \end{pmatrix}, \ 6N \times 6N \tag{10}$$

The computation of the hydrodynamic tensor is actually the most time consuming part of the simulations. The need for matrix diagonalization at each time step sets a practical limit to this method of about 100 diffusing particles.

2.2. TEST SIMULATION

This algorithm was implemented into a computer code using an object–oriented design. As a small test system we simulated the diffusion of 8 simplified positively charged cytochrome c molecules near a negatively charged membrane with embedded reaction centres of negative charge, see Figure 2.

This test simulation took about 50 minutes on a Pentium III 599 MHz PC with 128 MB Ram. With a parallel version of the program we will in future be able to simulate larger systems with more realistic interactions. In this way we will make contact with cell–simulation models such as Mcell (http://www.mcell.cnl.salk.edu) that aim at modelling subcompartments of biological cells. Mcell allows to simulate Brownian motion of thousands of particles by generating random walks using a Monte Carlo technique. In contrast to our method, Mcell does not compute forces between particles and with the surrounding cellular structures.

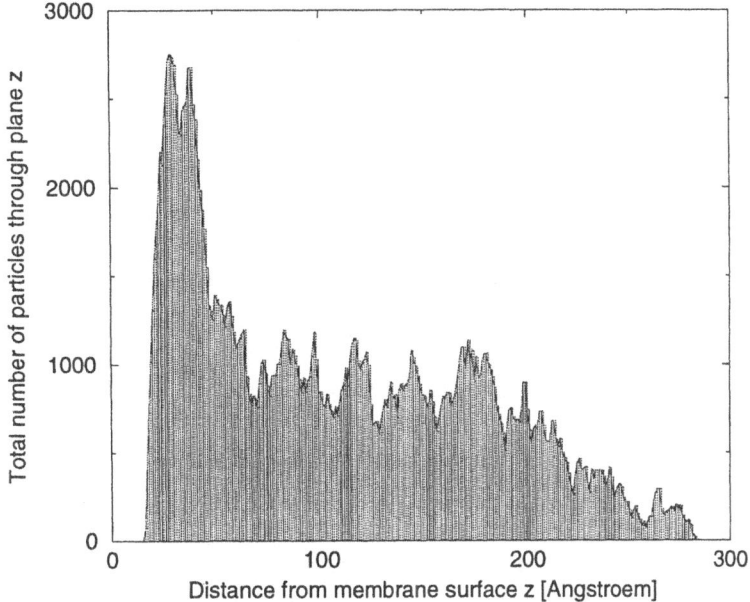

Figure 2. Distribution of the diffusing particles along the z–axis during the simulation. Since the membrane and cytochrome c's have opposite charges the probability to find cytochrome c's has a maximum near the surface. **Simulation parameters**: number of steps $N \simeq 30000$, iteration step $\Delta t = 10\ ps \rightarrow t_{total} \simeq 30\mu s$, Temperature $T = 298K$, Box side length $l = 30nm$, **Solvent parameters (water)**: Debye screening length $l_D = 1\ nm$, viscosity $\eta = 0.89\ 10^{-3}\ \frac{kg}{ms}$, dielectric constant $\epsilon = 78$, **Membrane parameters**: Surface Charge Density $\sigma = -\frac{0.277e}{(6.51\ nm)^2}$, Reaction centre modelled after X-ray structure 1PCR, radius $R = 3.1\ nm$, charge $q = -6\ e$ located $0.5\ nm$ above the membrane surface, **Cytochrome c parameters** [5]: 8 particles, radius $R = 1.66\ nm$, monopole: charge $q = +7.35\ e$ in the centre, dipole: charges $q = +1.725\ e$ and $q = -1.725\ e$ at $r = 1.6\ nm$ and $r = -1.6\ nm$, respectively, mass $m = 12.5\ kDa$.

3.   EFFICIENT SIMULATION OF PROTON TRANSFER BY Q–HOP MD

Another crucial step in photosynthesis and respiration is the creation of an external pH gradient across the biological membrane that is used by the protein $F_oATPase$ to synthesize ATP. The active transport of protons is carried out by large membrane–spanning proteins. A number of different theoretical approaches have previously been described to dynamically simulate proton transfer in systems ranging from the simplest systems such as an excess proton in bulk water to enzymatic reactions. The methods vary largely in complexity from classical models [6] over empirical valence bond (EVB) models [7, 8, 9] to Car–Parrinello simulations [10, 11]. The theoretical methods are becoming increasingly sophisticated and are becoming able to describe kinetic and thermodynamic aspects of individual proton transfer events in bulk solution and in macromolecules at high accuracy. So why did we feel it necessary to develop yet another technique?

First of all, some of these methods such as Car–Parrinello MD are computationally very expensive and are restricted to systems of small size covering time scales of 10–20 ps simulation time. However, transport of a single proton through a biological membrane protein may take as long as one millisecond for bacteriorhodopsin, one of the best studied proton pumps, and we therefore require the technique to be very efficient.

Secondly, protein residues may be essential parts of the proton-transporting hydrogen bond networks. The method should therefore not only work for bulk water, but be extendable to other chemical groups with reasonable effort.

Finally, we found it attractive to compute energy barriers and transfer probabilities at every simulation time step, because many clever techniques have been developed to speed up conformational sampling in molecular simulations. Once the relative barriers were known for various possible transfer paths, these barriers could in principle be manipulated to speed up the sampling.

To reach long simulation times for large systems, classical simulation techniques based on empirical force field methods are desirable. While MD simulations of biomolecular systems are commonly performed for a fixed protonation state, a simple way to include flexible protonation is to allow proton hopping between titratable sites. For example, Brickmann and co-workers [6] simulated an excess proton in liquid water. In this picture the dynamics of the system in a given protonation state is treated by classical MD simulation. Proton transfer from a proton donating amino acid or water molecule to an accepting one is treated by instantaneous proton hopping applying a simple distance criterion [6]. It would, however, be very difficult to find such distance criteria for complex systems with transfers taking place between a number of different chemical groups. As described below, we extended this approach by deriving simple fitting formulas to accurately compute the proton transfer probability between all possible donor–acceptor pairs.

3.1.   DERIVATION OF HOPPING PROBABILITIES

To calculate transfer rates under the electrostatic influence of environmental groups in an easy and fast manner, fit formulas were derived from quantum chemical studies on proton transfer reactions in small model systems [12, 13]. In a typical MD simulation starting with the system in a given protonation state, the distances $R(DA)_{ij}$ between all donor–acceptor pairs i and j are determined in spheres of a given radius around each donor atom i. For each such pair, the energy difference E_{12}^0 is calculated between the state where the proton is bound to the acceptor and the donor–bound state in vacuum. This is possible using a simple quadratic formula [13]

$$E_{12}^0 = \alpha + \beta\, R(\mathrm{DA}) + \gamma\, R(\mathrm{DA})^2 \quad , \tag{11}$$

where α, β and γ are parameters tabulated for all pairs of chemical groups present in the simulation, and $R(\mathrm{DA})$ is the only input value measured during the simulation.

Now, the electrostatic influence of the surrounding groups on the relative energies of donor and acceptor bound state needs to be determined. A sphere of pre–defined cut–off radius is drawn around the midpoint between donating and accepting atom of this donor–acceptor pair. The electrostatic interaction between donor–acceptor pair and the environmental groups in the sphere is calculated as

$$E_1^{env} = \sum_{i\in\{D-H,A\}} \sum_{j\in\{env\}} \frac{q_i q_j}{4\pi\varepsilon_0 |\mathbf{r_i} - \mathbf{r_j}|} \quad , \tag{12}$$

where $q_{i/j}$ and $\mathbf{r_{i/j}}$ are charges and coordinates of atoms i and j. E_1^{env} corresponds to the donor–bound state (protonated donor $D--H$, unprotonated acceptor A). To calculate the environmental influence on the putative acceptor bound state, a dummy situation is constructed where the donor is de–protonated and the acceptor is protonated, respectively. The point charges are modified according to the force–field charges of de–protonated donor D and protonated acceptor group $A-H$, and E_2^{env} is calculated as

$$E_2^{env} = \sum_{i\in\{D,H-A\}} \sum_{j\in\{env\}} \frac{q_i q_j}{4\pi\varepsilon_0 |\mathbf{r_i} - \mathbf{r_j}|} \quad . \tag{13}$$

So, the energy difference between donor and acceptor bound states with environmental influence E_{12} is

$$E_{12} = E_{12}^0 + E_2^{env} - E_1^{env} \quad . \tag{14}$$

Three different cases have to be distinguished depending on $R(\mathrm{DA})$ and E_{12} [14].

(a) For small $R(\mathrm{DA})$ and small E_{12} only a small or even no energy barrier $E_{\vec{B}}$ between donor and acceptor is present and the probability for transfer is high. This probability can be estimated by evaluating

$$p_{SE} = (0.5 \tanh(-K E_{12} + M) + 0.5)\frac{\Delta t}{10\,\text{fs}} \quad , \tag{15}$$

where $K = K(R(\text{DA}))$ and $M = M(R(\text{DA}))$ are easily calculated using tabulated parameters and $R(\text{DA})$ as input value. This formula was derived by studying the time evolution of wavepackets on the one–dimensional energy surface of the proton in the donor–acceptor system using the time-dependent Schrödinger equation [14].

(b) For large E_{12} and large $R(\text{DA})$ the energy barrier is so high that simple transition state theory is valid. In this case, the energy barrier $E_b^{\rightarrow}$ is estimated by

$$E_b^{\rightarrow} = S + T E_{12} + V (E_{12})^2 \tag{16}$$

$S = S(R(\text{DA}))$, $T = T(R(\text{DA}))$ and $V = V(R(\text{DA}))$ are easily calculated using tabulated parameters with $R(\text{DA})$ as input value, again. The calculated energy barrier is then used to estimate the zero–point energy [12]

$$\hbar\omega_1/2 = f\,exp(-g\,E_b^{\rightarrow}) + h \tag{17}$$

(f, g and h are tabulated values) and the enhancement of the transfer rate due to tunneling [14] to account for the quantum nature of the light proton.

$$\kappa(T) = \exp(P + Q E_M + R E_M^2) \tag{18}$$

(P, Q and R are tabulated and E_M is the minimum of forward and backward transfer barrier, $E_M = min(E_b^{\rightarrow}, E_b^{\leftarrow})$). Introducing $E_b^{\rightarrow}$, $\hbar\omega_1/2$ and $\kappa(T)$ into simple transition state theory gives the transfer probability in this regime

$$p_{TST} = \kappa(T)\frac{k_B T}{2\pi\hbar} \exp\left(-\frac{E_b^{\rightarrow} - \hbar\omega_1/2}{k_B T}\right)\Delta t \quad . \tag{19}$$

(c) For the intermediate regime a linear interpolation on logarithmic scale is used which bridges the gap between the limiting cases (a) and (b),

$$\begin{aligned}
\log_{10} p_{Gap} &= \log_{10} p_{SE}(E_{12}^L) + \frac{\log_{10} p_{TST}(E_{12}^R) - \log_{10} p_{SE}(E_{12}^L)}{E_{12}^R - E_{12}^L} \\
&\times (E_{12} - E_{12}^L),
\end{aligned} \tag{20}$$

where E_{12}^L and E_{12}^R are the estimated validity limits of the two approaches (a) and (b) [14].

Having calculated the transfer probability in one of the three ways a random number is computed and compared with the probability of proton transfer. If the random number is larger than the probability no hopping occurs. This comparison is done for every donor–acceptor pair. If no transfer process takes place at all, the simulation continues with the next MD–time steps in the same protonation state. If the random number is smaller than the probability, hopping takes place between donor and acceptor. The proton is instantaneously shifted from donating atom to accepting atom, and the simulation continues with MD–time steps in the altered protonation state.

3.2.   EXCESS PROTON IN A WATER BOX

Computational details.   An excess proton, i.e. H_3O^+, was simulated in a solvent box with 211 SPC water molecules and using periodic boundary conditions [15]. The system temperature was coupled to $T = 300$ K by a Berendsen thermostat and the system pressure to 10^5 Pa accordingly. The leap–frog algorithm was used with a time step of 2 fs to numerically solve Newton's equation of motion. The long–range Coulomb interactions were calculated up to a cut–off distance of 12 Å. Hopping was allowed every 10 fs. After a successful proton transfer, no back transfer was allowed for 20 fs. H_3O^+ was modelled with fixed bond length and flexible bond angles taken from the AMBER95 force field. 16 simulations were carried out over 0.5 ns.

Results.   The proton transfer rate is calculated as 0.47 ps^{-1} which is very close to the results of a quantum chemical study (0.50 ps^{-1} with centroid molecular dynamics [8, 9]) and is in the same order of experimental NMR results (~ 0.63 ps^-1).

To shed light on the microscopic structural details of proton transfer, the upper panel of Figure 3 displays the distances between hydronium oxygen and the oxygen atoms of the three closest neighbor molecules. The energy difference between the donor and acceptor bound state, E_{12}, is shown in the lower panel, respectively. The vertical lines mark the proton hopping events in this time window. The closest water molecule is usually at 2.4 Å to 2.6 Å distance of the donating oxygen. Most of the transfer events observed here happen at distances smaller than 2.5 Å. What determines hopping or non–hopping is the environmental configuration. In most hopping events shown here, the second and third closest water molecules are further away than on average. This leads to a destabilization of the hydronium ion due to weaker electrostatic interaction between the hydrogen bonded water molecules, not considering the acceptor. Thus, the transfer to the closest water molecule becomes en-

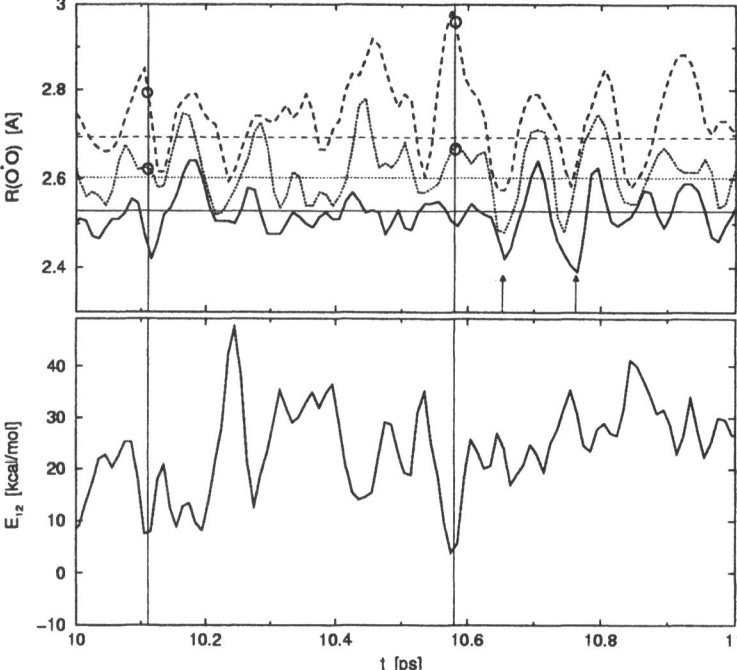

Figure 3. Upper panel: Distances between the hydronium oxygen and the oxygen atoms of the three closest neighbor molecules. Two separate transfer events are indicated by vertical lines, and the mean distances of the three closest molecules during this time frame are shown as horizontal lines. The vertical lines mark the transfer events and the dots the actual distances between donor and next water molecules at this moment. Lower panel: The electrostatic difference E_{12} between donor and next acceptor is shown in parallel.

ergetically more favorable. In the cases where all three water molecules are close to the donor (arrows), such destabilization of the hydronium ion does not occur, and the transfer is energetically unfavorable.

Another experimentally available observable is the diffusion coefficient of the excess proton. 16 simulation runs over 500 ps were performed of an excess proton in a box of 211 water molecules. The diffusion constant was computed using the Einstein relation

$$D = \left\langle \frac{|\mathbf{r}(t) - \mathbf{r}(0)|^2}{6t} \right\rangle \quad , \tag{21}$$

where the averaging was performed over the hydronium oxygen position $\mathbf{r}$ at different times t. The computed result of $9.0 \cdot 10^{-5} \pm 3.9 \cdot 10^{-5} \mathrm{cm}^2 \mathrm{s}^{-1}$ is in good agreement with the experimental value of $9.3 \cdot 10^{-5} \mathrm{cm}^2 \mathrm{s}^{-1}$

which is quite remarkable because the parametrization of the hopping probabilities was completely independent of any dynamic simulation.

4.　OUTLOOK

Computer simulations are a powerful tool nowadays that can often generate *novel insights* into biological processes. The computational techniques useful in biological systems span a wide range from quantum chemistry to colloidal sciences. Besides improvement of individual methods, methodological advances are required:

- to develop generally usable software packages for parallel architectures

- to interface different simulation methodologies, e.g. interface the Brownian dynamics description of diffusing particles with a finite element description of a lipid bilayer,

- in particular for the Q–HOP MD method presented find connection to analytical theories from chemical and statistical physics.

Acknowledgments

The multi protein project is supported by a grant of the Klaus Tschira Stiftung GmbH Heidelberg. The author also wishes to thank the organizers of the beautiful congress in Mexico City for the opportunity to participate at this very stimulating conference.

References

[1] D.J. Bicout and M.J. Field, *J. Phys. Chem.* **100** (1996) 2489.

[2] D.L. Ermak and J.A. McCammon, *J. Chem. Phys.* **69** (1978) 1352.

[3] E. Dickinson, S.A. Allison and J.A. McCammon, *J. Chem. Soc. Farad. Trans. 2* **81** (1985) 591.

[4] P. Mazur and W. van Saarloos, *Physica* **115A** (1982) 21.

[5] S.H. Northrup, in *Cytochrome C: A Multi disciplinary Approach*; (University Science Books, 1996) pp. 543.

[6] R. G. Schmidt and J. Brickmann, *Ber. Bunsenges. Phys. Chem.* **101** (1997) 1816.

[7] R. Vuilleumier and D. Borgis, *Chem. Phys. Lett.* **284** (1998) 71.

[8] U. W. Schmitt and G. A. Voth, *J. Phys. Chem.* **B102** (1998) 5547.

[9] U. W. Schmitt and G. A. Voth, *J. Chem. Phys.* **111** (1999) 9361.

[10] D. Marx, M. E. Tuckerman, J. Hutter, and M. Parrinello, *Nature* **397** (1999) 601.

[11] M. E. Tuckerman, D. Marx, M. L. Klein, and M. Parrinello, *Science* **275** (1997) 817.

[12] M.A. Lill and V. Helms, *J. Chem. Phys.* **114** (2001) 1125.

[13] M.A. Lill, M.C. Hutter and V. Helms, *J. Phys. Chem.* **A104** (2000) 8283.

[14] M.A. Lill and V. Helms, *J. Chem. Phys.* **114** (2001) 7985.

[15] M.A. Lill and V. Helms, *J. Chem. Phys.* **114** (2001) 7993.

FROM NEURON TO BRAIN: STATISTICAL PHYSICS OF THE NERVOUS SYSTEM

Paul H.E. Tiesinga

Sloan–Swartz Center for Theoretical Neurobiology and Computional Neurobiology Lab,
Salk Institute
10010 N. Torrey Pines Rd, La Jolla, CA 92037, USA
tiesinga@salk.edu

Abstract   A common experimental neuroscience protocol is to record the single neuron activity in response to repeated stimulus presentations and analyze how this activity encodes for stimulus properties. Neurons are embedded in a large network and their response properties depend on the dynamical state of the network. I discuss how brain chemicals — neuromodulators — can dynamically change the coherence and frequency content of network activity; how the network coherence affects the neuronal response to current injection and how this can form the neural correlate for an important computation — gain modulation — that the brain performs.

Keywords: neuron, synchronized oscillations, neural coding, gain modulation

1.   INTRODUCTION

The brain is a massively interconnected network of neurons that generates spatiotemporally ordered activity on multiple time and length scales. Individual neurons produce all–or–none electric events known as action potentials or spikes. A basic belief in neuroscience is that neurons encode information using action potentials. The fundamental question is how information is encoded, transmitted, processed, and decoded in cortical networks. To make progress, one needs to understand the spiking dynamics of local cortical circuits and how to link their activity to the coherent macroscopic activity detected in electroencephalograms, functional magnetic resonance imaging, and magnetoencephalograms. Biologically informed application of quantitative techniques based on nonlinear dynamics, complex systems theory, statistical physics and information theory are likely to yield important contributions to solving this class of problems.

Developments in Mathematical and Experimental Physics,
Volume B: Statistical Physics and Beyond
Edited by Macias *et al.*, Kluwer Academic/Plenum Publishers, 2003

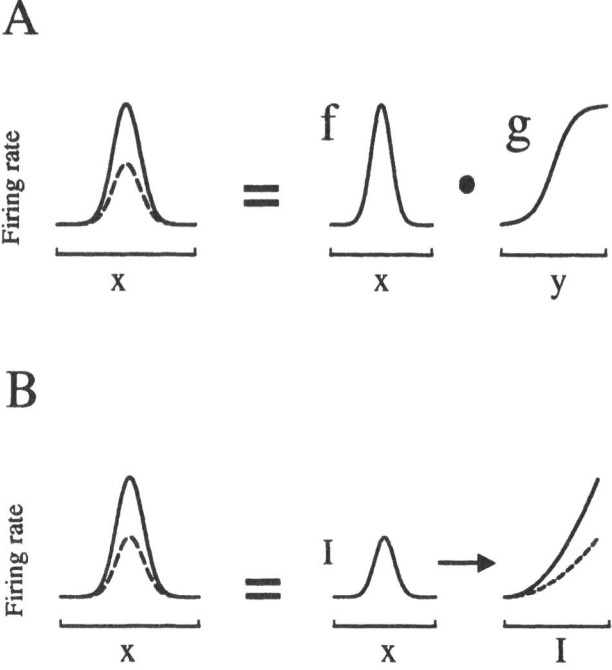

Figure 1.   A possible single neuron correlate of gain modulation. (A) Firing rate tuning curves as a function of x for two different values of y (solid and dotted line, respectively). The neural response can be expressed as a product of $f(x)$, the tuning curve for x, and $g(y)$, the gain as a function of y. (B) Stimulus property x causes a tuned current I into the neuron. The resulting firing rate is given by an f–I characteristic that depends on y (the solid and dotted line represent two y values).

In this work I review one particular aspect of local circuit dynamics. A typical neuron receives both input from the network it is embedded in, as well as feedforward input from, for instance, sensory areas. I will discuss how the network state affects the neuron's response to feedforward inputs and, furthermore, what purpose modulation of network activity could potentially subserve.

A common experimental paradigm for studying neural coding is to present a visual stimulus to an animal and record spike trains from single neurons in the appropriate sensory area. A key response quantity is the spike count — the number of spikes emitted by the neuron in a fixed time interval after stimulus presentation. Stimulus characteristics are then systematically varied in order to assess how they are encoded by the neuron using the spike count.

The mean neuronal firing rate (equal to the spike count normalized for the duration of the measurement) elicited by a stimulus presentation can often be written as a product of stimulus attributes. For instance, the response of neurons in visual cortex to oriented bars is tuned for the orientation x (dotted line in Fig. 1A). When the brightness of the stimulus, y, is increased a similar but rescaled tuning curve is obtained (solid line in Fig. 1A). Hence, the response is a product of the orientation tuning $f(x)$ and the brightness–induced gain modulation $g(y)$ (Fig. 1A).

Gain modulation is a general computational paradigm in the nervous system and it is observed in many cortical areas (reviewed in Refs. [1, 2, 3]). Attention could also gain-modulate orientation tuning curves [4, 5]. Recently, neural correlates of attention were studied experimentally using awake behaving monkeys [6, 7, 8]. Exactly the same two sensory stimuli were presented to the monkey under two conditions: when it had to pay attention to one of the stimuli to perform a task for which it was rewarded; and when it did *not* have to pay attention to that stimulus in order to obtain its reward. During this task, spike trains from one or more neuron(s) in sensory areas were recorded. The feedforward sensory input that the recorded neuron received should be the same under both conditions and any difference in neuronal response between the conditions had to be due to the network state. In one experiment the response of multiple neurons in the secondary somatosensory cortex to tactile stimulation of the finger tips was recorded [6]. When the animal had to pay attention to the tactile stimulus the number of coincident spikes between two simultaneously recorded neurons increased significantly. This effect was independent of changes in firing rate. In the other experiment [7], two visual stimuli were presented to the animal. The first stimulus was in the receptive field of the neuron that was recorded from, whereas the second stimulus was outside the receptive field. The local field potential — an electrical signal that represents the mean activity of a group of neighboring neurons — was recorded simultaneously with the single neuron activity. When the animal focused attention on the stimulus in the receptive field two changes were observed compared with when it payed attention to the stimulus outside this receptive field: (1) the local field potential power in the gamma–frequency band (30-80 Hz) had increased and (2) the coherence of the single neuron discharge with the local field potential had increased. However, the firing rate had not changed significantly.

Both these experiments show that attention can modulate the state of cortical networks, in particular, neural synchrony. Here I explore how neural synchrony affects the gain of a neuron's response. A key assumption is that stimulus attribute x causes a tuned depolarizing current $I(x)$ into the neuron (Fig. 1B). The resulting firing rate is given by the firing rate versus current curve ($f–I$) whose gain depends on stimulus attribute y. Hence, the problem is to determine how the gain of the $f–I$ depends on the amount of synchrony.

The rest of the paper is organized as follows. First, the effect of incoherent network activity on the $f–I$ is studied. Second, the generation of coherent activity by networks is discussed. Then, the effect of coherent network activity on the $f–I$ is investigated. Finally, gain modulation by coherent and incoherent network activity is compared.

2.   RESULTS

2.1.   GAIN MODULATION BY "BALANCED" SYNAPTIC INPUTS.

The impact of network activity on the response properties of neurons was investigated using model simulations. Full details are given in Ref. [9] (related work is in Refs. [10, 11, 12]). Briefly, a Hodgkin–Huxley–type model neuron was driven by a constant current I and a combination of excitatory and inhibitory shot noises with rates f_e and f_i, respectively. The shot noises represented the network inputs. Each pulse generated a conductance pulse that decayed exponentially in time. The conductance generated a membrane current, I_{syn}, given in Eq. (A.1) of the Appendix. Parameters were chosen to mimick GABA$_A$ inhibitory and AMPA excitatory synaptic inputs [13].

A typical voltage response is shown in Fig. 2A. Network background activity was $f_e = f_i = 750$, but during the time interval indicated by the bar it was reduced to $f_e = f_i = 500$. Action potentials are visible as sharp voltage deflections that cross 0 mV. This computer experiment was repeated 50 times with different shot noise realizations (trials). The spike times were plotted as ticks in a rastergram, each row was a different trial (Fig. 2Ba). The spike time histogram (STH, Fig. 2Bb) was computed by counting the number of spikes in 200 ms bins (normalized to yield spikes per trial per second). During the period of reduced network activity there was an immediate increase in output firing rate.

The amount of injected depolarizing current was varied for different levels of background activity. The output firing rate increased with the level of depolarizing current. Furthermore, the rate of increase — gain — of the resulting $f–I$ depended strongly on network activity: a higher background activity led to a lower gain.

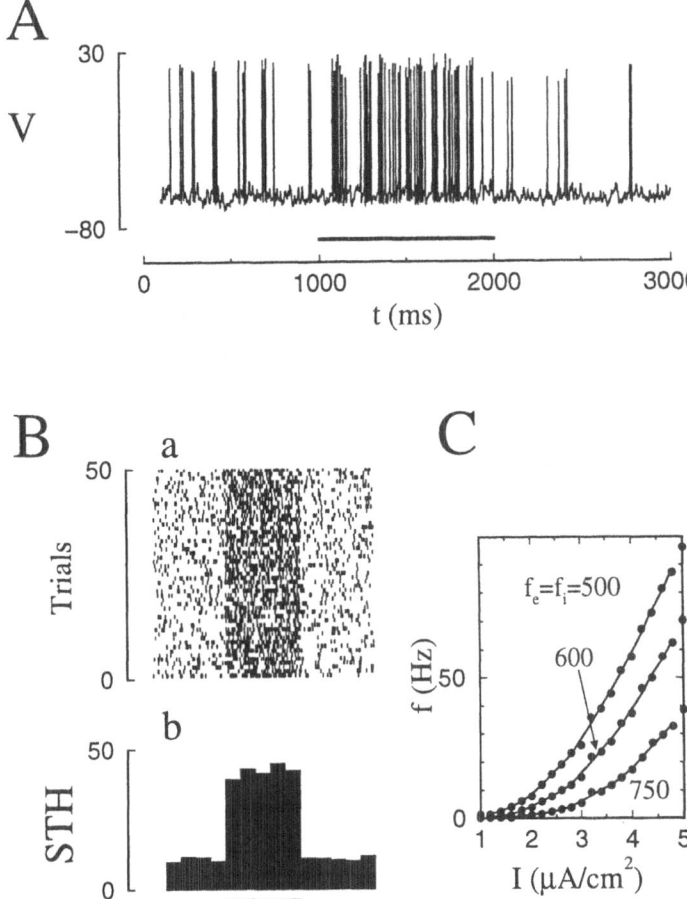

Figure 2. Gain modulation by "balanced" synaptic inputs. The background synaptic activity was reduced from $f_i = f_e = 750$ Hz to 500 Hz between $t = 1000$ ms and 2000 ms (indicated by the bar in A and Bb). (A) Voltage (in mV) versus time during the first trial and (B) (a) rastergram and (b) spike time histogram (STH) for 50 trials is shown. (C) The mean firing rate as a function of injected current I for different levels of background synaptic activity as is indicated in the graph. Averages were obtained over 20 s.

What is the significance of this result? The depolarizing current is the feedforward signal coming from sensory areas. The neural response to the signal, I, thus depends on the state of the network the neuron is embedded in. The firing rate recorded *in vivo* therefore does not uniquely represent a particular input signal. However, as pointed out by Chance & Abbott [14], this mechanism can form the neural substrate for gain modulation.

The type of network activity the neuron received is sometimes called balanced [11, 15, 16, 17] (or fluctuation–dominated [9]). The idea is that the *average* effect of the excitatory and inhibitory inputs is approximately the same but opposite in sign, that is, balanced. The mean drive remains below spiking threshold and spiking is caused by fluctuations. The fluctuations can be varied by increasing the network activity in such a way that it remains balanced. For incoherent inputs, Poisson spike trains with a time–independent rate, there are only two parameters that can be varied: the rates f_i and f_e (under the assumption that the unitary amplitude and decay time constant of the pulses remain constant). However, the synaptic drive yields a mean conductance, Eq. (A.2) and a mean current, Eq. (A.3) (see Appendix). Only one of these can be held constant. Hence, one can have either have a balanced "current" [9] or "conductance" drive [14].

2.2. COHERENT ACTIVITY OF NEURONAL NETWORKS

Neuronal networks spontaneously generate synchronized oscillations in different frequency bands. For instance, the human electroencephalogram (EEG) contains multiple synchronous rhythms that reflect spatially and temporally ordered activity [18, 19]. Delta rhythms (0.5 – 2 Hz) are involved in signal detection and decision making [20]; theta oscillations (4 – 12 Hz) are related to cognitive processing [21]; and gamma oscillations (30 – 80 Hz) can be found in a variety of structures under different behavioral conditions [18, 22, 23]. The neural mechanisms for these correlated changes in EEG frequency content are unknown, but they are likely to involve a class of brain chemicals known as neuromodulators [24, 25, 26]. For instance, in vivo studies in cats show that the level of the neuromodulator acetylcholine in the cortex and hippocampus (a brain area involved in memory processes) is reduced during delta oscillations, and can be up to four times higher during theta activity [27].

The in vitro hippocampal preparation has provided a simplified framework, which allows the mechanisms underlying theta, delta and gamma rhythms, and the transitions between the rhythms to be studied. Activation of the CA3 region in hippocampus, using the agonist carbachol (a substance with similar effects as acetylcholine), results in synchronous population activity in the delta [28, 29], theta [30, 29, 31, 32, 33], and gamma [29, 34] bands. The carbachol concentration in the slice determines which of the three oscillations will occur or predominate [29].

The biophysical effects of carbachol have been studied extensively. Each neuron has a diverse set of voltage–gated ionic channels in its membrane. The density and types of channels determines a neuron's dynamics. Carbachol blocks several channels in a concentration dependent manner [35, 36]. This makes neurons more excitable, unmasking subthreshold membrane potential oscillations in the theta–frequency range [37, 29]. In addition, carbachol reduces the average strength of excitatory synaptic coupling between neurons [38, 39, 40]. Therefore, carbachol changes the intrinsic neuronal dynamics and the network dynamics, which has made it difficult to dissect these two contributions solely by experimental methods. Recently my colleagues and I determined using model simulations under what conditions, and by what mechanisms, the hippocampal area CA3 can support the CCH-δ, CCH-θ, and CCH-γ oscillations [41]. In particular, we investigated how carbachol induced transitions between these states in a concentration dependent manner. These results show that even a single neuromodulator can exert a powerful control on the coherence and frequency content of network activity. I will now focus on gamma oscillations in networks of interneurons [42, 43, 44], since, as mentioned in the introduction, gamma–band activity is thought to be important for attention. Attention was associated with increased synchrony of the modulatory input but the mean firing rate did not significantly increase. I explored under what conditions synchrony of a coupled network could be varied independently of mean firing rate. The result for an all–to–all coupled inhibitory network is shown in Fig. 3A. The neurons in this network are connected by inhibitory synapses. Hence, when a particular neuron becomes more active, it *reduces* the activity of the neurons it is connected to.

A depolarizing current pulse of 500 ms duration was applied to all neurons at $t = 1000\ ms$. The pulse represented the effects of, for instance, the release of acetylcholine. Initially, the network was asynchronous, the spike time histogram did not have any peaks and the mean firing rate in the network was approximately 8 Hz. It took approximately 300 ms for the network to synchronize, at that time sharp peaks appeared in the spike time histogram. The firing rate only increased marginally to 9 Hz. It is, of course, not known whether the effects of attention can be represented by a depolarizing current or whether the attention-

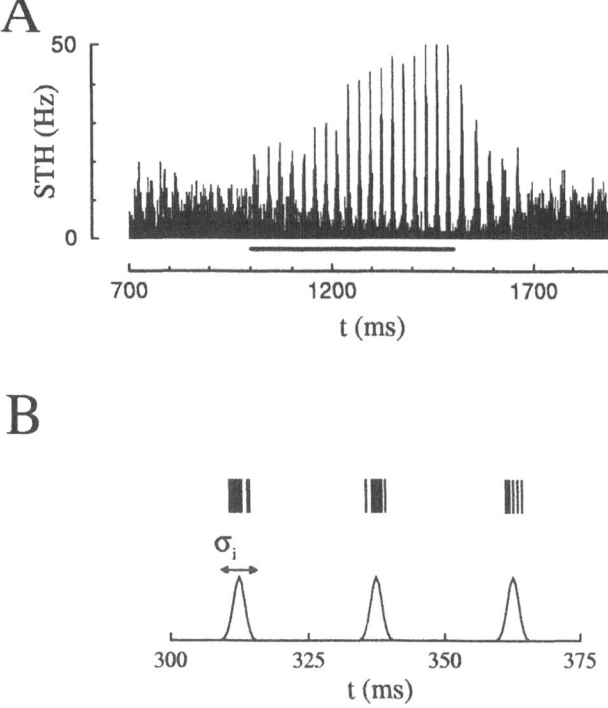

Figure 3. (A) The inhibitory network synchronized in response to a brief depolarizing current pulse (indicated by the bar) to 25% of the neurons. (B) Coherent network discharge was modelled as (top) an inhomogeneous Poisson spike train with (bottom) a spike time probability density consisting of a periodic sequence of Gaussians of width σ_i. The probability density was normalized so that the mean rate was f_i.

related networks have an all–to–all architecture. However, the results are a proof of principle that it is possible to modulate neural synchrony independently of mean network activity.

Synchronous network activity can be modelled in a less computationally intensive way as is shown in Fig. 3B. The network spike time histogram consisted of a sequence of Gaussian functions with width σ_i. Input spike times were generated as a Poisson process from the STH [44, 45, 46]. The degree of neural synchrony was correlated with the width, σ_i, of peaks in the STH. Low values of σ_i corresponded to high synchrony and high values of σ_i corresponded to low synchrony. As before, each spike generated unitary conductance pulses of amplitude Δs_i and a decay time τ_i (see Appendix for details).

2.3. GAIN MODULATION BY NEURAL SYNCHRONY.

The neuron received two types of inputs, a feedforward projection that was modelled as a constant depolarizing current and a modulatory input representing, for instance, the effects of attention. The modulatory input was modelled as an synchronous inhibitory synaptic drive. The effects of attention were modelled by reducing σ_i from 8 ms to 2 ms during a 500 ms long pulse (indicated by the bar in Fig. 4A). Without attention the neuron did not fire, but as soon as input synchrony was increased it started firing. The attentional effect was reversible: as soon as the synchrony was reduced to its original value the neuron stopped firing.

There were two sources of variability in the synaptic inputs: jitter in the position of the Gaussian peaks in the spike time density for the Poisson process (Fig. 3B, bottom) and the particular realization of the Poisson spike train given a particular spike time density (Fig. 3B, top). To assess the effect of the variability on the neuronal response the simulation was repeated 500 times with independent realizations (Fig. 4B). The transient increase of synchrony reliably led to an increased firing rate (Fig. 4Ba). The resulting spike time histogram during the period of increased input synchrony was flat. This means that although there were signs of output periodicity during a given trial (Fig. 4A), the timing across different trials was not maintained. Hence, in this case, the precise timing of spikes on a given trial does not carry stimulus information. Note, however, that the spike time histogram obtained from an *ensemble* of neurons receiving the same synaptic drive will be peaked.

The firing rate versus current characteristic was determined for three values of the synchrony parameter σ_i (Fig. 4C). As before, the firing rate increased with the level of depolarizing current. However, the rate of increase was higher for more synchronous inputs, that is, lower values of σ_i. At first instance, this may seem counterintuitive: A more synchronous inhibitory drive results in a more effective inhibition yet it yields a higher firing rate. Just after the arrival of the input pulses the synaptic conductance reaches its highest value. Over a time period on the order of the synaptic decay constant, τ_i, the conductance decays to its lowest value. The synaptic drive thus has two components, the phasic component, quantified as the difference between maximum and minimum conductance value, and the tonic component equal to the minimum conductance value. The neuron can only fire when inhibition has sufficiently decayed. Whether or not a neuron can fire is determined by the value of the tonic component. Higher synchrony — lower

σ_i — leads to a higher phasic and a lower tonic component. The parameters of the simulations were chosen such that for an asynchronous drive, $\sigma_i = 10$ ms, the tonic component was so large that the neuron did not spike, whereas for a synchronous drive it could fire. An alternative viewpoint is as follows. The neuron can be approximately considered as a threshold device. The parameters were tuned so that the neuron was below threshold and could only fire when voltage fluctuations drove it across threshold. A higher phasic component leads to larger voltage fluctuations, hence a higher firing rate. This hypothesis was confirmed by comparing the variance of voltage fluctuations with the firing rate when varying σ_i (results not shown).

Gain modulation with synchrony is only possible when the firing rate was actually modulated by σ_i. The conditions under which this occurred were determined. Two parameters were important: the frequency, f_{net}, of the synchronous inhibitory barrages (equal to the inverse of the distance between consecutive Gaussian peaks in Fig. 3B, the default value was $f_{net} = 40$ Hz) and the synaptic decay time constant τ_i. A measure for the maximal gain is the firing rate for $\sigma_i = 1$ divided by the firing rate for $\sigma_i = 10$. This ratio had an optimal value as a function of f_{net}. For $\tau_i = 10$ ms — the default value used here to model GABA$_A$ synaptic inputs — the position of the optimum was in the gamma frequency range. The position of the optimum depended on τ_i: it shifted to higher frequencies as τ_i was decreased. Full details will be given elsewhere [47, 48]

3.   DISCUSSION

Gain modulation is an important computational mechanism for representing two or more stimulus attributes in a neuron's firing rate. Here two possible mechanisms for implementing gain modulation were presented. They work by modulating the response to feedforward sensory inputs that were modelled as a depolarizing current. The mechanisms were studied by calculating the firing rate versus current characteristic (f–I). The first mechanism is based on balanced synaptic inputs and was proposed by Chance & Abbott [14]. The gain of the f–I was reduced by proportionally increasing both the excitatory and inhibitory activity. An alternative mechanism proposed here is based on modulating synchrony. The gain of the f–I increased with synchrony. I also showed using model simulations that network synchrony could easily be manipulated using neuromodulators. The gain modulation corresponds to a multiplicative rescaling of the $f--I$. This is sometimes hard to assess from the $f-I$ curves themselves. However, the sensitivity of the

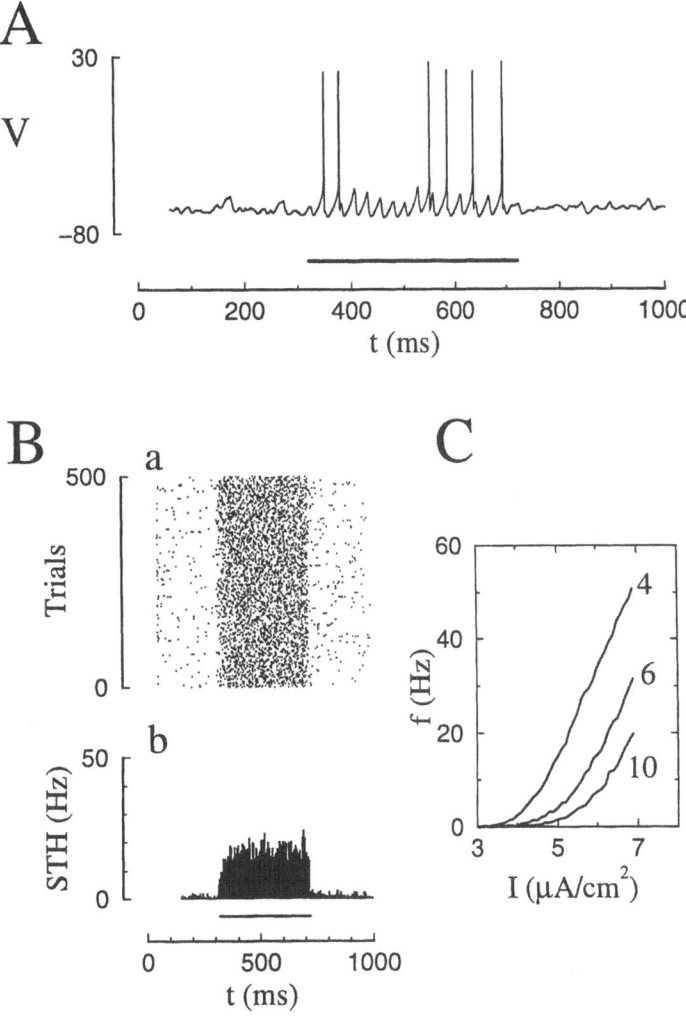

Figure 4. Gain modulation by synchrony. The synchony was increased between $t = 270$ ms and 520 ms by decreasing σ_i from 8 ms to 2 ms. (A) Voltage (in mV) versus time during one trial, (B) (a) rastergram and (b) STH for 500 trials. (C) f–I characteristic for different values of σ_i, as is labeled in the graph.

firing rate to current changes, df/dI, plotted versus I or f yields a better test of multiplicative rescaling (L.F. Abbott, personal communication).

The underlying biophysics was similar for both mechanisms: it uses the fact that neurons are threshold devices. The membrane voltage of the neuron is on average subthreshold and spikes are induced by fluctuations. The gain is adjusted by changing the variance of the fluctuations while the neuron remains subthreshold. For balanced synaptic inputs, the fluctuations are modulated by covarying excitatory and inhibitory input rates at a fixed ratio. Increasing input rates changes both the mean synaptic conductance as well as the mean current. The ratio between f_i and f_e can be chosen to keep either the mean current or mean conductance constant, but not both at the same time. Hence, the distance to threshold may also change during manipulations of the balanced synaptic input. For gain modulation with synchrony, the mean conductance always remained the same, and the mean driving current changed very little for the parameters considered here (results not shown). Hence, synchrony works by directly affecting voltage fluctuations, without changing the means of the drive.

The dynamics of single neurons is often studied in the *in vitro* preparation. A slice of cortical tissue is kept "alive" in a solution resembling corticospinal fluid. A depolarizing current is injected into the neuron using an intracellular electrode and the voltage response is recorded. The firing rate of the neuron — number of action potentials per second — is then determined as a function of the amplitude, I, of the injected current. *In vitro* neurons do not receive the synaptic drive due to network activity. Recently, experimental techniques have been developed to study "*in vivo*" dynamics of neurons using the *in vitro* slice preparation. The network activity is injected into the neuron using so-called dynamic clamp [49]. Some of the modelling results reported here were confirmed using this technique [47]. Hence, both mechanisms could potentially subserve gain modulation in cortical tissue. Recent experiments have shown that modulating network coherence is an important component of attention, but a conclusive *in vivo* experiment explicitly linking attentional gain modulation to synchrony is still lacking. Furthermore, an important theoretical issue is how to generalize gain modulation of one neuron to that of a interacting network of neurons.

Acknowledgments

The results reported in this proceedings are based on work done in collaboration with Jean–Marc Fellous, Jorge V. José, Emilio Salinas and Terry Sejnowski. I thank the Sloan–Swartz Center for Theoretical Neurobiology for financial support. Part of the calculation was performed at Northeastern University High Performance computer center. I thank the organizers of the "1st Mexican Meeting on Mathematical and Experimental Physics" for their hospitality during my visit to Mexico city.

Appendix

The synaptic drive injected into the model neuron was a sum of inhibitory and excitatory conductances,

$$I_{syn} = g_e s_e(t)(V - E_e) + g_i s_i(t)(V - E_i). \tag{A.1}$$

Where the conductance was $g_e = g_i = 1.0\,mS/cm^2$, and the reversal potentials were $E_e = 0\,mV$ and $E_i = -75\,mV$, for the excitatory (AMPA) and inhibitory (GABA$_A$) synapses, respectively. Unitary excitatory (inhibitory) postsynaptic potentials were modelled as quantal conductance increases, $\Delta s_e = 0.02$ ($\Delta s_i = 0.05$), in the synaptic kinetic variable $s_e(t)$ ($s_i(t)$). The conductance pulses in $s_e(t)$ and $s_i(t)$ decayed exponentially in time with a time constant $\tau_e = 2$ ms ($\tau_i = 10$ ms). The postsynaptic potentials were independent and formed either a homogeneous or an inhomogeneous (see Fig. 3B) Poisson process with average rates f_e and f_i, respectively.

The mean of the total synaptic conductance was,

$$\langle g_{syn} \rangle = g_e \langle s_e \rangle + g_i \langle s_i \rangle, \tag{A.2}$$

and the mean of the driving current I_{syn} was

$$\langle I_{syn} \rangle = \beta_e \langle s_e \rangle + \beta_i \langle s_i \rangle, \tag{A.3}$$

with $\beta_e = g_e(E_e - V_{rest})$, and $\langle s_e \rangle = \tau_e f_e \Delta s_e$ (with similar expressions for the inhibitory part); V_{rest} was the resting membrane potential. The last expression is only valid for weak synaptic drives, unlike the ones used here, see Ref. [9] for details.

References

[1] E. Salinas and P. Thier, *Neuron* **27** (2001) 15.

[2] E. Salinas and T. Sejnowski, *Nat Rev Neurosci* **2** (2001) 539.

[3] E. Salinas and T. Sejnowski, *Neuroscientist* **7** (2001) 430.

[4] C. McAdams and J. Maunsell, *J. Neurosci.* **19** (1999) 431.

[5] S. Treue and J. Martinez Trujillo, *Nature* **399** (1999) 575.

[6] P. Steinmetz *et al.*, *Nature* **404** (2000) 187.

[7] P. Fries, J. Reynolds, A. Rorie, and R. Desimone, *Science* **291** (2001) 1560.

[8] S. Treue, *Trends Neurosci.* **24** (2001) 295.

[9] P. Tiesinga, J. José, and T. Sejnowski, *Phys. Rev.* **E62** (2000) 8413.

[10] D. Pare and A. Destexhe, *J. Neurophysiol.* **81** (1999) 1531.

[11] E. Salinas and T. Sejnowski, *J. Neurosci.* **20** (2000) 6193.

[12] A. Destexhe, M. Rudolph, J.-M. Fellous, and T. Sejnowski, *Neuroscience* **107** (2001) 13.

[13] G. Shepherd, *The synaptic organization of the brain* (Oxford University Press, New York, 1998).

[14] F. Chance and L. Abbott, *Soc. Neurosci. Abstr.* **26** (2000) 1064.

[15] M. Shadlen and W. Newsome, *J. Neurosci.* **18** (1998) 3870.

[16] S. Song, K. Miller, and L. Abbott, *Nat. Neurosci.* **3** (2000) 919.

[17] A. Burkitt, *Biol. Cybern.* **85** (2001) 247.

[18] E. Basar, C. Basar-Eroglu, S. Karakas, and M. Schurmann, *Neurosci. Lett.* **259** (1999) 165.

[19] A. Glass and R. Riding, *Biol. Psychol.* **51** (1999) 23.

[20] C. Basar–Eroglu, E. Basar, T. Demiralp, and M. Schurmann, *Int. J. Psychophysiol.* **13** (1992) 161.

[21] O. Vinogradova, *Prog. Neurobiol.* **45** (1995) 523.

[22] R. Ritz and T. Sejnowski, *Curr. Opin. Neurobiol.* **7** (1997) 536.

[23] W. Singer and C. Gray, *Annu. Rev. Neurosci.* **18** (1995) 555.

[24] D. McCormick, *Prog. Neurobiol.* **39** (1992) 337.

[25] M. Stewart and S. Fox, *Trends Neurosci.* **13** (1990) 163.

[26] L. Leung, *Neurosci. Biobehav. Rev.* **22** (1998) 275.

[27] F. Marrosu *et al.*, *Eur. J. Neurosci.* **7** (1995) 358.

[28] J. Boguszewicz, B. Skrajny, J. Kohli, and S. Roth, *Can. J. Physiol. Pharmacol.* **74** (1996) 1322.

[29] J.-M. Fellous and T. Sejnowski, *Hippocampus* **10** (2000) 187.

[30] B. Bland, L. Colom, J. Konopacki, and S. Roth, *Brain Res.* **447** (1988) 364.

[31] J. Konopacki, M. MacIver, B. Bland, and S. Roth, *Brain Res.* **405** (1987) 196.

[32] B. MacVicar and F. Tse, *J. Physiol. (Lond)* **417** (1989) 197.

[33] J. Williams and J. Kauer, *J. Neurophysiol.* **78** (1997) 2631.

[34] A. Fisahn, F. Pike, E. Buhl, and O. Paulsen, *Nature* **394** (1998) 186.

[35] D. Madison, B. Lancaster, and R. Nicoll, *J. Neurosci.* **7** (1987) 733.

[36] D. McCormick, *Trends Neurosci.* **12** (1989) 215.

[37] L. Leung and C. Yim, *Brain Res.* **553** (1991) 261.

[38] M. Hasselmo, *Behav. Brain Res.* **67** (1995) 1.

[39] M. Hasselmo, E. Schnell, and E. Barkai, *J. Neurosci.* **15** (1995) 5249.

[40] M. Patil and M. Hasselmo, *J. Neurophysiol.* **81** (1999) 2103.

[41] P. Tiesinga, J.-M. Fellous, J. José, and T. Sejnowski, *Hippocampus* **11** (2001) 251.

[42] X.-J. Wang and G. Buzsáki, *J. Neurosci.* **16** (1996) 6402.

[43] J. White *et al.*, *J. Comput. Neurosci.* **5** (1998) 5.

[44] P. Tiesinga and J. José, *Network* **11** (2000) 1.

[45] P. Tiesinga, J.-M. Fellous, J. José, and T. Sejnowski, *Network* **13** (2002) 41.

[46] P. Tiesinga, J.-M. Fellous, J. José, and T. Sejnowski, *Neurocomputing* **38-40** (2001) 397.

[47] J. José *et al.*, *Soc. Neurosci. Abstr.* 722.11 (2001).

[48] P. Tiesinga *et al.*, in preparation (2002).

[49] A. Sharp, M. O'Neil, L. Abbott, E. Marder, *Trends Neurosci.* **16** (1993) 389.

ELASTIC VIBRATIONAL OF NANOPARTICLES

E. Haro–Poniatowski, J. Hernández–Rosas, M. Picquart
Universidad Autónoma Metropolitana Iztapalapa, Departamento de Física,
Apdo postal 55-534, Col. Vicentina, 09340 México D.F., México
haro@xanum.uam.mx

M. Jouanne, J. F. Morhange, M. Kanehisa
Université Pierre et Marie Curie, LMDH, UMR 7603
4 Place Jussieu, 75252 Paris Cedex 05, France
mj@ccr.jussieu.fr

Abstract   Low frequency acoustic vibrations have been observed by Raman spectroscopy in a wide variety of nano-metric objects in various conditions. A simple elastic model has been used in many cases to give account for the observed results. In this work results obtained on bismuth nanocrystals embedded in an amorphous matrix, free standing zirconia powders and a monoclonal antibody molecule are analyzed and compared.

Keywords: Nanoparticles, low frequency vibrations, Raman spectroscopy.

1.   INTRODUCTION

The vibrations of a sphere under stress–free boundary conditions at the surface were computed more than a century ago by Lamb and Love using theory of elasticity [1, 2]. This calculation is similar to compute the eigenfunctions in the case of an hydrogen atom; one finds that the spectrum is discrete and depends on three discrete numbers, n, l, m corresponding to the azimuthal, angular and radial orders. This theory has been successfully applied to the breathing mode of the earth observed for the first time during an earthquake in 1960 in Chile [3]. Actually these modes have been detected in planets of the solar system as well. At the other extreme, the possible observation by Raman spectroscopy of the normal modes of vibration of nanosized objects was first suggested by De Gennes and Papoular in the particular case of globular proteins

Developments in Mathematical and Experimental Physics,
Volume B: Statistical Physics and Beyond
Edited by Macias *et al.*, Kluwer Academic/Plenum Publishers, 2003   113

in 1969 [4]. During the seventies several accounts of broad bands in the low frequency range of the Raman spectra of some proteins were reported. The simple elastic model was used to explain the observed results. However due to the lack of experimental data concerning the elastic properties of biological objects data corresponding to polymeric materials was used instead [5].

Raman scattering by confined acoustical phonons in nanoparticles has been studied since the early eighties. It has successfully been applied to small particles of different materials. In semiconductor doped glasses such as CuCl, CdS, CdSe, CdS_xSe_{1-x}, Ge [6, 7, 8]. In metallic clusters embedded in insulating materials which exhibit enhanced dielectric and optical properties [9, 10, 11]. In free standing TiO_2 powders [12] and in nanoparticles suspended in a liquid solution [13]. A very important parameter to determine in these materials is the size of the nanoparticles. Low frequency Raman Scattering has been proposed as an alternative technique to measure size. The size effects in the low frequency region of the Raman spectrum have been put in evidence unambiguously in various systems. For example in the case of ZrO_2 powders calibrated sizes have been prepared and their corresponding Raman spectrum characterized [14, 15]. However this has not been the case for the biological nano–objects for obvious reasons, in this case the size cannot be controlled. In fact other models involving hydrogen bond have been invoked to explain the observed low frequency bands [16]. Measurements by inelastic neutron scattering have been used to support these assumptions. In this work, a large size range of nanopowders of ZrO_2, of Bismuth nanoparticles having an oblate ellipsoidal shape embedded in an amorphous Germanium matrix were investigated by Raman spectroscopy. Finally, we present also the results concerning a protein, the IgG2a inmunoglobulin.

2.   ELASTIC VIBRATIONS OF A SPHERE

The equation describing the propagation of an acoustic wave in an elastic medium is [2]:

$$\mu\nabla^2\vec{u} + (\lambda+\mu)\vec{\nabla}(\vec{\nabla}\cdot\vec{u}) = \rho\frac{\partial^2\vec{u}}{\partial t^2} \tag{1}$$

its solution satisfies:

$$\vec{u} = \vec{\nabla}\Phi + \vec{\nabla}\times\vec{A} \tag{2}$$

Φ and $\vec{A}$ are solutions of the following wave equations respectively

$$C_L^2\nabla^2\Phi = \frac{\partial^2\Phi}{\partial t^2} \tag{3}$$

$$C_T^2 \nabla^2 \vec{A} = \frac{\partial^2 \vec{A}}{\partial t^2} \tag{4}$$

Longitudinal and transverse sound velocities, C_L and C_T, can be defined in terms of the Lamé coefficients and the density:

$$C_L = \sqrt{\frac{\lambda + 2\mu}{\rho}} \tag{5}$$

$$C_T = \sqrt{\frac{\mu}{\rho}} \tag{6}$$

Assuming harmonic time dependence, equation involving Φ becomes the Helmholtz scalar wave equation and the equation involving $\overrightarrow{A}$ becomes the Helmholtz vector wave equation.

The boundary conditions are obtained by setting up the stress tensor in order to obtain the eigenvalues;

$$\sigma_{ik} = \lambda u_{ii}\delta_{ik} + 2\mu u_{ik} \tag{7}$$

where u_{ik} is the strain tensor given by:

$$u_{ii} = \frac{\partial}{\partial x_i}\frac{u_i}{\sqrt{h_{ii}}} + \frac{1}{h_{ii}}\sum_{k=1}^{3}\frac{\partial h_{ii}}{x_k}\frac{u_k}{\sqrt{h_{kk}}} \tag{8}$$

$$u_{ik} = \frac{1}{2\sqrt{h_{ii}h_{kk}}}\left[h_{ii}\frac{\partial}{\partial x_k}\left(\frac{u_i}{\sqrt{h_{ii}}}\right) + h_{kk}\frac{\partial}{\partial x_i}\left(\frac{u_k}{\sqrt{h_{kk}}}\right)\right] \tag{9}$$

Where the trace of the strain tensor is the divergence of the displacement vector $\vec{u}$ and represents the relative change in volume of the body. h_{ii} represents the scale factors in the corresponding coordinate system.

In the case of free standing oscillations the stress tensor is zero at the boundary, furthermore the solutions must be finite at the origin. An alternative way to solve equation (1) is to decompose the displacement vector into its longitudinal and transversal parts:

$$\vec{u} = \vec{u}_L + \vec{u}_{T1} + \vec{u}_{T2} \tag{10}$$

In this case the solutions for the spherical case can be written as:

$$\begin{aligned}
\vec{u}_L &= \frac{1}{k_L}\vec{\nabla}(Y_l^m(\theta,\varphi)j_l(k_L\vec{r})) \\
&= \frac{1}{k_L}(\vec{P}_{ml}(\theta,\varphi)\frac{d}{dr}j_l(k_L r) + \sqrt{l(l+1)}\vec{B}_{ml}(\theta,\varphi)j_l(k_L r)) \tag{11}
\end{aligned}$$

$$\vec{u}_{T1} = \vec{\nabla} \times (\vec{r} Y_{ml}(\theta, \varphi) j_l(k_T r)) = \sqrt{l(l+1)} \vec{C}_{ml}(\theta, \varphi) j_l(k_L r) \qquad (12)$$

$$
\begin{aligned}
\vec{u}_{T2} &= \frac{1}{k_T} \vec{\nabla} \times \vec{u}_{T1} \qquad\qquad\qquad\qquad\qquad (13) \\
&= \frac{1}{k_T r} \left[l(l+1) \vec{P}_{ml}(\theta, \varphi) j_l(k_T r) + \sqrt{l(l+1)} \vec{B}_{ml}(\theta, \varphi) \frac{d}{dr} (r j_l(k_T r)) \right]
\end{aligned}
$$

where $Y_l^m(\theta, \varphi)$ and $j_l(k_L r)$ are the spherical harmonics and the spherical Bessel functions, respectively. $\vec{P}_{ml}(\theta, \varphi)$, $\vec{B}_{ml}(\theta, \varphi)$ and $\vec{C}_{ml}(\theta, \varphi)$ are the vector spherical harmonics.

Another way to write equation (7) considering the present boundary conditions in a sphere of radius R is by expressing the strain tensor produced by the three types of displacements $\vec{u}_L$, $\vec{u}_{T1}$ and $\vec{u}_{T2}$ as a function of the forces. The forces are related to the strain (stress) tensor by $T_i = \sigma_{ik} n_k$ where n_k is the unit vector in the direction of the free surface vibrating sphere. The forces are therefore zero for free standing oscillations and it is necessary to put (11), (12), (13) in (7) in order to obtain a system where the determinant must be zero.

With these conditions one finds for the torsional modes:

$$\frac{\partial}{\partial a} \left(\frac{j_l(\alpha)}{\alpha} \right) = 0 \qquad (14)$$

where l ($l \geq 1$) is the angular momentum and the spheroidal modes:

$$
2 \left[\beta^2 + (l-1)(l+2) \{ \beta \frac{j_{l+1}(\beta)}{j_l(\beta)} - (l+1) \} \right] \alpha \frac{j_{l+1}(\alpha)}{j_l(\alpha)} - \frac{1}{2} \beta^4
$$
$$
+ (l-1)(2l+1)\beta^2 + \{ \beta^2 - 2l(l-1)(l+2) \} \beta \frac{j_{l+1}(\beta)}{j_l(\beta)} = 0 \qquad (15)
$$

where α and β are given by:

$$\alpha = \frac{\omega R}{C_L} \qquad (16)$$

$$\beta = \frac{\omega R}{C_T} \qquad (17)$$

The spheroidal modes equation reduces to spherically symmetric compressional modes for l=0. For l different from zero the deformations will be nonspherical, for example for l=1 the sphere evolves from a prolate to an oblate spheroid.

3.   RAMAN SPECTROSCOPY
EXPERIMENTS

All spectra were first reduced by the Bose–Einstein factor. Such a procedure is mandatory because of the low frequency and large widths of the observed modes. Experimental conditions details are given in ref. [17]. To analyze the experimental data below 250 cm^{-1}, the reduced spectra were fitted to a sum of various damped oscillators [17]. For all the samples, the frequencies of all the oscillators are deduced from this fit.

3.1.   ZIRCONIUM OXIDE

Zirconium oxide (ZrO_2) nanosized powders present various phases depending on the calcination temperature (T_c) [14, 15]. For T_c lower than 310°C, the Raman spectrum is typical of an amorphous material and, as usual, is an image of the density of states of the material. Around 360°C, bands characteristic of the tetragonal phase of Zirconia are observed. As the temperature increases, there is a progressive splitting of the bands associated with the appearance and growth of monoclinic phase. For this phase (space group C_{2h5}), 18 modes are expected to be Raman active, most of which are effectively observed. While the frequency shift of the peaks is primarily due to the increasing size of the ZrO_2 crystallites [18], the mixture of the metastable tetragonal phase with the monoclinic phase obscures the effect. It is thus difficult to extract information concerning the size of the nanocrystals from the optical phonon domain of the Raman spectrum.

An expanded view of the frequency region under 200 cm^{-1} is shown on Figure 1 which presents the Raman spectra of the powders for different calcination temperatures (a) 330°C, (b) 360°C, (c) 400°C, (d) 500°C, (e) 700°C, (f) 1000°C. As it is obvious in this figure, the low frequency feature (under 80 cm^{-1}) can be decomposed into two broad bands. When the temperature of calcination increases, these two bands shift towards lower frequencies.

In the case of zirconia 6 oscillators were used. The size of the nanoparticles was obtained by X-rays diffractography. For the lowest calcination temperatures (200, 250 and 300°C), because of the amorphous nature of the sample, broad peaks are only observable making impossible to extract a size information from the X-rays diffractograms. The corresponding size values were obtained by extrapolation. We have compared the experimental values of the ω_{0i} deduced from the fit to a calculation of the frequencies of the first two Raman active modes [19] of nanoparticles modelled as spheres. The frequencies (cm^{-1}) of these modes are

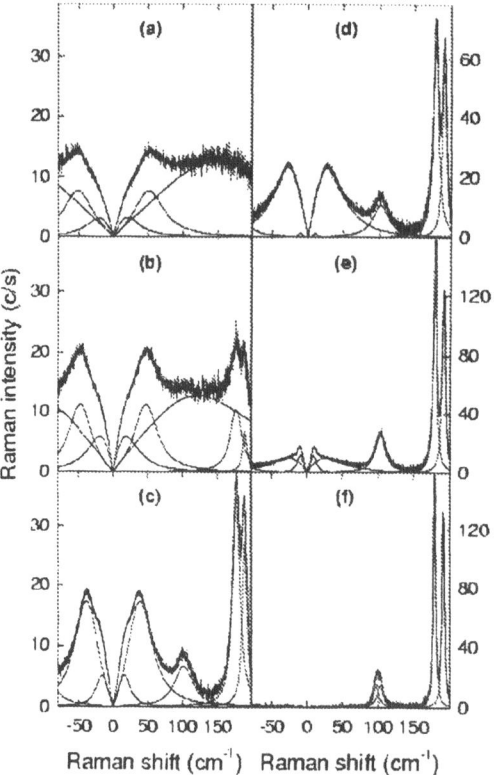

Figure 1. Low frequency Raman spectrum of the zirconia samples annealed at 330, 360, 400, 500, 700 and 1000°C

inversely proportional to their radius, according to the expression

$$\omega = (\frac{X}{\pi})\frac{C_T}{30}\frac{1}{D} \tag{18}$$

where C_T is the transverse velocity of sound in m/s, D the diameter or the particle in nm and $X = \pi\beta_{ln}$ and β_{ln} is the nth-root of $J_l(\pi\beta) = 0$ [1]. To achieve this calculation, the sound velocities (longitudinal and transverse) of the material are required. We have used the elastic constants E = 2.44×10^{12} dyne/cm^2 (Young modulus) and $\mu = 0.97 \times 10^{12}$ dyne/cm^2 extrapolated for zero porosity monoclinic polycrystalline ZrO_2 to calculate the Lamé constant $\lambda = 1.03 \times 10^{12}$ dyne/cm^2 and using the zero porosity density of 5.71 g/cm^3 of reference [20], we obtain for the longitudinal and transverse sound velocities: C_L = 7 212 m/s and C_T = 4 122 m/s. These values have been used for the three phases of ZrO_2 involved in the range of temperature where the two low frequency modes are observed.

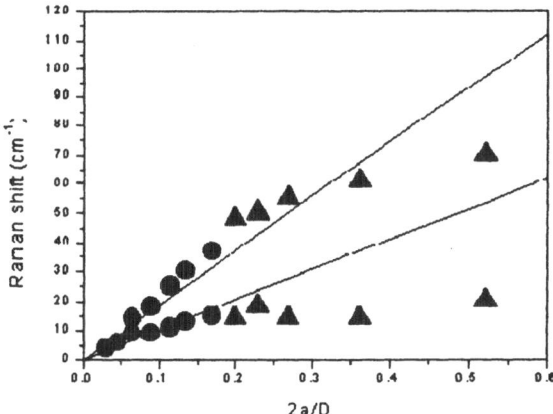

Figure 2. Raman shift as a function of normalized size. The two straight lines represent the calculation using Eq. 18

The calculated frequencies versus the inverse of the radius for the two possible Raman active fundamental spheroidal modes $l = 0$ and $l = 2$, (the $l = 2$ of torsional mode being Raman forbidden) are plotted on figure 2, using these sound velocities. A reasonable fit is observed between the lines and the experimental points for the largest grain sizes, considering the crude approximation on the sound velocities and the non spherical shape of the nanoparticles. For the smallest sizes (low calcination temperatures), the agreement is poorer as one would expect since the approximation of the elastic properties of continuous media used for the calculations is no longer valid. This behavior is similar to the bending of the acoustical branches of the dispersion curves for large wavevectors.

3.2. BISMUTH

In figure 3 the low frequency Raman spectra of the samples as a function of the Bi particle size are presented: (a) 2.31 nm, (b) 2.51 nm, (c) 7.63 nm, (d) 17.55 nm and (e) 23.41nm respectively. The spectrum of a monocrystalline bulk material in parallel and perpendicular polarizations was added (f). On the monocrystalline sample, the two main peaks observed are the well known modes E_g ($70\ cm^{-l}$) and A_{1g} ($97\ cm^{-l}$) of the Bismuth. A small shoulder on the low frequency of the E_g peak as well as a broad background extending over more than $100\ cm^{-l}$ can also be observed on the same spectrum. This continuum is peaked at around $22\ cm^{-l}$ and cannot be related to stray light. An interpretation in terms of various combinations of phonons, eg A_{1g}(G)-E_g(G) at $27\ cm^{-l}$, 2

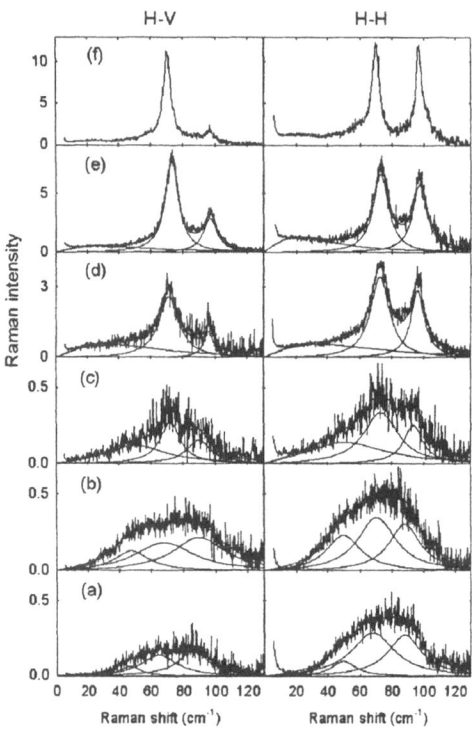

Figure 3.   Low frequency Raman spectrum of Bi nanoparticles as a function of size: (a) 2.31 nm, (b) 2.51 nm, (c) 7.63 nm, (d) 17.55 nm, (e) 23.41 nm and (f) monocrystalline Bi.

TA(X) at 45 cm^{-1} and 2 LA(X) at 67 cm^{-1} [21] could be responsible for this effect. On the contrary, for the thin films, the low frequency band is still present at low temperature. The continuum band is more difficult to interpret. It is observed on the spectrum of bulk Bismuth but undergoes some shift when the size of the particles decreases. In the bulk this band could be interpreted in terms of higher order combination of phonons while for the thin films another interpretation is needed. A possible origin is the vibrational modes of small sized particles [7, 19] which can be regarded as confined acoustic modes. The intensity of this mode decreases when the particle size increases as the confinement becomes less effective. To consider this possibility, we have calculated the highest frequency of the two fundamental spheroidal Raman active modes of a clamped sphere of bismuth, using the classical theory of Lamb [1]. The clamping of the sphere assumes that the amorphous Germanium matrix is much more rigid than the Bismuth particles. For this calculation, the sound velocities were taken from reference [22] to be $C_L = 2.18 \times 10^3$ m/s and $C_T = 1.1 \times 10^3$ m/s for polycrystalline Bismuth. The length of the ellipsoid axis normal to the surface has been estimated to be smaller than 2 nm for all the samples [23]. Similarly to the case discuss above for the zirconia nanoparticles when their sizes increases, the spherical approximation holds. However for small particles the experimental frequencies fall well above the theoretical prediction [24].

3.3.   PROTEINS

In Fig. 4 the low frequency Raman spectrum of IgG2a antibody is presented. A structured band with a maximum at 80-90 cm^{-1} is clearly observable. This band has also a shoulder on its low frequency side. The upper figure is the uncorrected spectrum, the corresponding reduced spectrum [17] is shown in the lower figure.

The IgG2a antibody is presented in Fig.5. A more realistic model for inmunoglobulins would be to consider vibrations of a rod. Deformation of a rod consists of three modes, stretching, twisting, and bending [25]. For a rod of finite length l, solution of the elastic wave equation becomes much more complicated and no analytic solution is available : we must have recourse to numerical methods. There is, however, an exception to this. For torsional waves, the eigenfrequency is given analytically by

$$\omega = C_T \sqrt{(\frac{\pi \beta_{2n}}{r})^2 + (\frac{m\pi}{l})^2}, \pi = 0, 1, 2, ...; m = 1, 2, 3, ... \qquad (19)$$

$\pi\beta_{2n}$, is the n–th root of the equation $J_2(\pi\beta) = 0$ where J_2 is the Bessel function of order 2 $\beta_{20} = 0$, $\beta_{21} = 1.6348$, $\beta_{22} = 2.6792$, ... Eq. (19) is valid for torsional motion. For other modes, although no analytic expression is available, we may use (19) as a rough estimate provided we replace the shear velocity by an appropriate velocity c. Thus we have

$$\omega = c \sqrt{(\frac{\pi \beta_{2n}}{r})^2 + (\frac{m\pi}{l})^2}, \qquad (20)$$

For the velocity, we use the value c = 2450 m s^{-1} determined in a previous paper [26]. As it can be seen in Fig. 5, the IgG2a antibody can be considered as constituted of small rods of the light chain size. The length is l = 80 $\overset{\circ}{A}$ and the radius r = 10 $\overset{\circ}{A}$, approximately [27]. As we have reported previously, the lowest modes n = 0, m = 1 have frequencies too small to be observed by Raman scattering because of the relatively large l. The modes with n = 1 can be observable because r is smaller. With the above values, the n = 1, m = 0 mode as a frequency of : $\omega = 66.7$ cm^{-1}

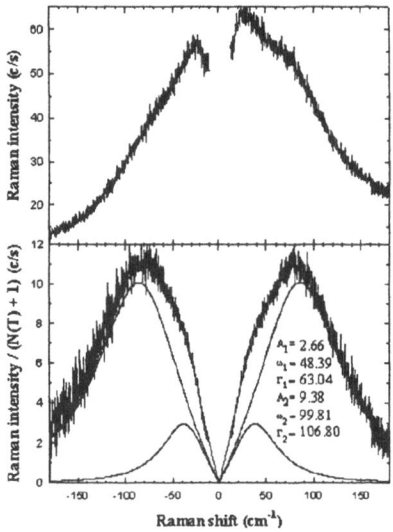

Figure 4.   Low frequency Raman shift of IgG2a.

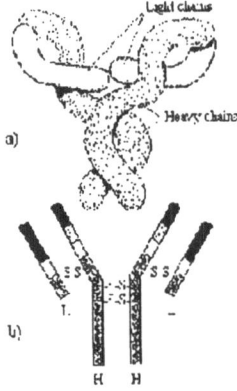

Figure 5.   Schematic representation of IgG2a.

The experimental values found after deconvolution of the low frequency band are: $48.4\ cm^{-1}$ y $99.8\ cm^{-1}$

We can also consider the antibody as constituted of 3 bigger cylindrical envelopes around the three lobes, as suggested by Rajan et al. [28], of $l = 80\ \AA$ and $r = 24\ \AA$, approximately. In this case the frequency would be of : $\omega = 27.8\ cm^{-1}$

The effect of the vibrations involving also the length of the rod is to increase the high frequency side of these bands. The agreement between calculated values and experimental is satisfactory if we consider that the rods are not perfect and that the radius is not constant.

4. CONCLUSION

From the results presented in this work one can establish unambiguously a variation of the Raman low–frequency feature as a function of size in the case of inorganic nanoparticles. For biological nanostructures the lack of calibrated sizes limits the possibility of finding such a relation. Further experiments in various biological systems with different sizes and shapes need to be made.

Acknowledgments

The present work has been partially supported by the Consejo Nacional de Ciencia y Tecnología (CONACyT) of México and the Centre National de la Recherche Scientifique (CNRS) of France.

References

[1] H. Lamb, *Proc. of the London Math. Soc.* **13** (1882) 189.

[2] A.E.H. Love, in: *A Treatise on the Mathematical Theory of Elasticity*, 4th ed., Dover, (Cambridge University Press, 1927) section 198.

[3] P. Lognonné *Bull. Soc. Fr. Phys.* **101** (1995) 3.

[4] P. G. de Gennes and M. Papoular, in: *Polarisation, Matière et Rayonnement. Volume in Honor of Alfred Kastler*, eds. J. Sahade, G. McCluskey, and Y. Kondo, Paris, (Presses Universitaires de France, 1969) pp. 243.

[5] W. L. Peticolas, in: *Enzymology 61, Enzyme Structure, Part H*, (Academic Press Inc, 1979) pp. 425.

[6] B. Champagnon, A. Boukenter, E. Duval, C. Mai, G. Vigier and E. Rodek, *J. Cryst. Solids* **94** (1987) 216.

[7] B. Champagnon, B. Andrianasolo and E. Duval, *Mat. Sci. Eng. B* **9** (1991) 417.

[8] M. Fuji, Y. Kanzawa, S. Hayashi and K. Yamamoto, *Phys. Rev.* **B54** (1996) R8373.

[9] G. Mariotto, M. Montagna, G. Viliani, E. duval, S. Lefrant, E. Rezpka and C. Mai, *Europhys. Lett.* **6** (1988) 239.

[10] M. Fuji, T. Nagareda, S. Hayashi and K. Yamamoto, *Phys. Rev.* **B44** (1996) 6243.

[11] M. Fuji, T. Nagareda, S. Hayashi and K. Yamamoto, *J. Phys. Soc. Jpn.* **61** (1992) 754.

[12] M. Gotic, M. Ivanda, S. Popovic, S. Music, A. Sekuli, A. Turkovic and K. Furic, *J. Raman Spectros.* **28** (1997) 555.

[13] P.C. Painter, L.E.Mosher and C. Rhoads, *Biopolymers* **21** (1982) 1469.

[14] E. Anastassakis, B. Papanicolaou and I. M. Asher, *J. Phys. Chem. Solids* **36** (1975) 667.

[15] A. Feinberg and C.H. Perry, *J. Phys. Chem. Solids* **42** (1981) 513.

[16] S. Melchionna and A. Desideri, *Phys. Rev. E* **60** (1999) 4664.

[17] M. Jouanne, J.F. Morhange, M. Kanehisa, E. Haro-Poniatowski, G.A. Fuentes, E. Torres and E. Hernández-Tellez, *Phys. Rev. B* **64** (2001) 5404.

[18] L. Fangxin, Y. Jinlong and Z. Tianpeng, *Phys. Rev. B* **55** (1997) 8847.

[19] E. Duval, *Phys. Rev. B* **46** (1992) 5795.

[20] C. F. Smith and W. B. Crandall, *J. Am. Ceram. Soc.* **47** (1964) 624.

[21] R.E. Macfarlane, in: *The Physics of Semimetal and Narrow Gap Semiconductors*,ed.D.L. Carter and R.T. Bate, Oxford, (Pergamon Press, 1971) pp 28.

[22] *The Handbook of Physical Quantities*, Boca Raton, (CRC Press, 1977).

[23] R. Serna, T. Missana, C. N. Afonso, J. M. Ballesteros, A. K. PetfordLong and R. C. Doole, *Appl. Phys.* **A66** (1998) 43.

[24] E. Haro–Poniatowski, M. Jouanne, M. Kanehisa, R. Serna and C.N. Afonso, *Phys. Rev.* **B60** (1999) 10080.

[25] L.L. Landau and E.M. Lifshitz, in: *Theory of Elasticity*, Oxford, (Pergamon Press, 1953).

[26] M. Picquart, E. Haro-Poniatowski , J.F. Morhange, M. Jouanne, and M. Kanehisa, *Biopolymers* **53** (2000) 342.

[27] L.J. Larson,K.W. Hasel and A. McPherson , *Biochemistry* **36** (1997) 1581.

[28] S.S. Rajan, K.R. Ely, E.E. Abola, M.K. Wood, P.M. Colman, R.J. Athay and A.B. Edmunson *Mol. Inmunology* **20** (1983) 787.

VIBRATIONAL SPECTROSCOPY OF BIOLOGICAL MEMBRANE MODELS

Michel Picquart

Universidad Autónoma Metropolitana Iztapalapa, Departamento de Física,

Apdo postal 55-534, Col. Vicentina, 09340 México D.F.

mp@xanum.uam.mx

Abstract   The purpose of this paper is to outline in the field of biological science the kinds of problems to which vibrational spectroscopy (infrared absorption and Raman scattering) can provide solutions. Emphasis is given here to the use of vibrational spectra in the study of molecular structures of biomembranes and interactions of biologically important materials. The main aspect will concern the changes observed in the lipid spectrum when host molecules are interacting: vitamin E, antibiotic (sodium lasalocid) and in the water molecules.

Keywords: Raman, FTIR, bilayer and, phospholipids.

1.   INTRODUCTION

The knowledge of membrane structure is important to understand all the processes that exist in the membrane function. The fluid mosaic model [1] describes the molecular arrangement that constitutes the biomembranes: a fluid lipid bilayer in which are immersed totally or partially proteins separates the inside from the outside. While the individual components, proteins and lipids, are constituted of covalent bonds, their association with each other is largely governed by the hydrophobic effect. [2]

Many of the properties of cell membranes are reflections of their lipid content. These lipids are diverse in structural details, but they have a common structural feature. They are amphiphilic molecules with a polar head and a non–polar acyl chain (Fig. 1). They are polymorphic and can exist in a variety of organized structure, especially when hydrated. This polymorphism depends not only on the structure of the lipid and on its degree of hydration but also on temperature, pressure, ionic strength and pH. In biomembranes, the structure is more complicated because

Developments in Mathematical and Experimental Physics,
Volume B: Statistical Physics and Beyond
Edited by Macias *et al.*, Kluwer Academic/Plenum Publishers, 2003

125

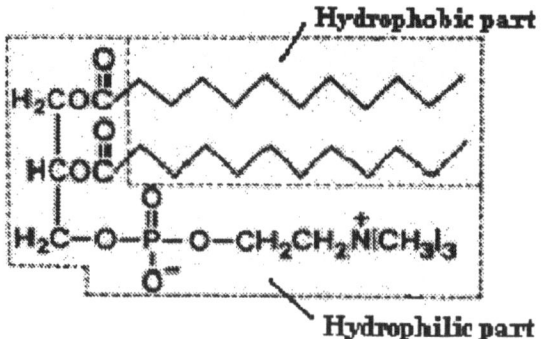

Figure 1.   Schematic representation of a phospholipid (phosphatidylcholine).

they do not contain only phospholipids but also proteins, glycoproteins, caroten-oids, and other molecules. Unilamellar vesicles (liposomes) or multilamellar vesicles (MLVs) are frequently used as models for biomembranes. They are roughly spherical particles enclosed by a lipid bilayer with an inner compartment that contains water and ions or hydrophilic compounds.

The number and types of phases formed by phospholipids, and thus their thermotropic phase behavior, depend strongly on its degree of hydration. Increasing water content results in a progressive decrease in the main phase transition temperature (T_m) until a certain water content is reached, where no further decrease in T_m is noted. This indicates that the progressive adsorption of water molecules decreases the strength of the interaction of adjacent lipid molecules in the bilayer. [3] The kinds of information obtained from the vibrational spectrum of bi-

Table 1.   Main bands observed in model membranes by Raman scattering and FTIR absorption and used in the analysis of the spectrum.

Bands	Frequency range (cm^{-1})
O-H and N-H stretching	3100 - 4000
C-H stretching	2800 - 3300
C=O stretching	1550 - 1870
CH_2 scissors	1410 - 1470
C-O-C stretching	1000 - 1300
PO_2^- stretching	1100 - 1250
C-C stretching	850 - 1200

ological molecules are the same as for smaller molecules, namely, the frequencies of intramolecular or intermolecular vibrations in the range 10 to 4000 cm^{-1}, the intensities of the spectral bands, and the polarization characteristics of the absorbed IR radiation or scattered Raman radiation. [4]

Many biological molecules are macromolecules and have a large number of fundamental vibrations (3N-6, where N is the number of atoms for a non-linear molecule and 3N-5 when linear). The vibrational spectrum is complicated, containing many overlapping bands from different molecular constituents. It is impossible to identify completely all vibrations for a biological molecule. Nevertheless, vibrations of a particular submolecular group of atoms appear with characteristic group frequencies (Table 1).

2. LIPID SPECTROSCOPY

Raman spectra are often studied in the 2750 - 3050 cm^{-1} and 1000 - 1200 cm^{-1} ranges (Fig. 2). The first region includes the contribution of the methyl and methylene CH stretching modes, and provides information on the perturbations arising in the hydrophobic core of the membrane and on the chain packing. The main bands are observed at 2850 and 2880 cm^{-1}, the symmetrical and antisymmetrical CH_2 stretching modes respectively. When melting occurs, the antisymmetrical methylene CH stretching mode intensity decreases in relation to the symmetrical one. The R=I_{2880}/I_{2850} ratio is sensitive to intra- and interchain order. The higher this ratio, the more ordered the hydrocarbon chains. [5]

The r=I_{2935}/I_{2880} ratio is also highly sensitive to intra- and interchain interactions, but it also reflects the mobility of the methyl end groups of hydrocarbon chains, as the $\nu_s(CH_3)$ vibration is also involved in this ratio. The smaller this ratio, the higher the conformational order of the lipid hydrocarbon chains, and the smaller the mobility of the methyl end groups. [6] The 1000 - 1200 cm^{-1} region includes the antisymmetrical (1060 cm^{-1}) and symmetrical (1130 cm^{-1}) C-C stretching modes (skeletal modes) and provides information on the intrachain disorder with the band at 1098 cm^{-1} representative of gauche bonds. The peak height ratios I_{1069}/I_{1098} or I_{1128}/I_{1098} therefore reflect the trans/gauche isomerisation along the lipid chain. The FTIR spectroscopy, and the analysis of the stretching modes of carboxyl groups (1650 - 1800 cm^{-1}), phosphate and $N^+(CH_3)_3$ groups (900 - 1300 cm^{-1}) of dipalmitoyl-phosphatidylcholine (DPPC) allow us to characterize the perturbations induced by host molecules in the polar and interfacial regions of the membrane. Bands arising from the interfacial and polar region provide important information concerning bilayer hydration. In this analysis, the ester C=O band of DPPC is particularly useful because it is composed

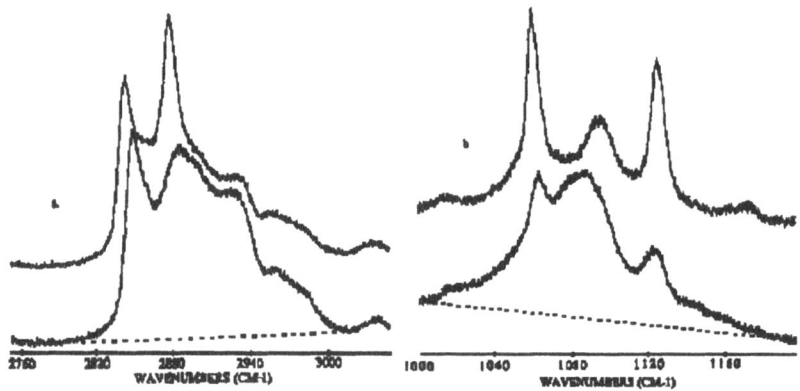

Figure 2.   Raman spectra in the CH (a) and CC (b) stretching regions at 10°C (lower) and 45°C of a DPPC aqueous solution

of two overlapping bands near 1740 and 1727 cm^{-1}. Many studies have been done to assign these two vibrations. [7, 8, 9] The higher–frequency component was assigned to free C=O groups of both acyl chains, and the low-frequency component to both monohydrated C=O groups but this interpretation is questioned and although hydration phenomenon cannot be excluded, differences in conformation of the polar head can be mentioned to explain changes in the $\nu_{C=O}$ region.

Hydration of DPPC [6, 10, 11, 12] with mechanical agitation in excess water produces a suspension of "onion-like" MLVs that undergo a sharp order - disorder transition at 41°C (midpoint of the phase transition). A small discontinuity is always observed near 35°C (the so-called pre-transition). This transition involves a change from the lamellar gel ($L_{\beta'}$) to the so-called ripple ($P_{\beta'}$) phase. The main transition is from the ripple to the fluid liquid crystalline (L_α) phase.

2.1.   LIPID - VITAMIN E

Figure 3a,b presents the temperature profiles for DPPC and DPPC / alpha-tocopherol ($\alpha - T$) multilayers, at different concentrations. Spectrum of pure $\alpha - T$ in the C-H stretching region shows that a band of the phytyl chain of $\alpha - T$ at 2935 cm^{-1} is very intense, perturbing this ratio. To prevent this influence correction of the intensity has been done [13]. When $\alpha - -T$ is added, the transition temperature decreases as the concentration increases and the transition region broadens (decrease of the transition cooperativity). The onset of the transition temperature is reduced as the concentration of $\alpha - -T$ increases, and

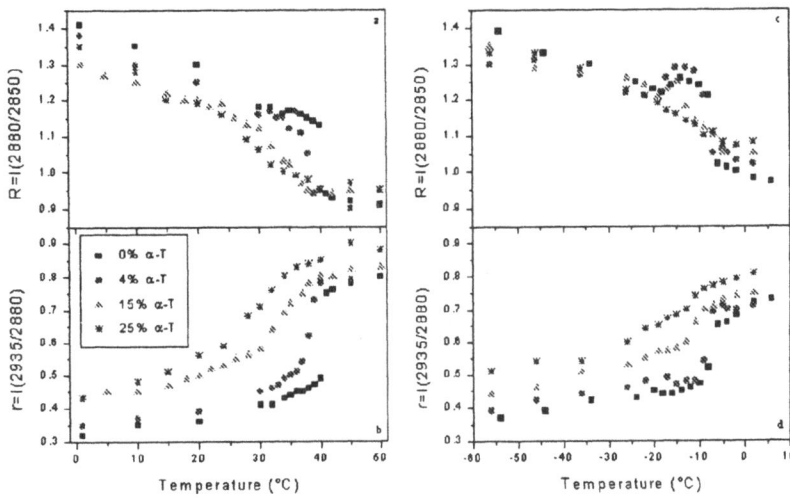

Figure 3. Temperature profiles constructed with the R and r ratios of DPPC (a, b) and OPPC (c, d) for increasing concentrations of α–tocopherol

the pre–transition vanished. The temperature profiles obtained with oleoylpalmitoyl–phosphatidyl–choline (OPPC) have many common features with that obtained with DPPC (Fig. 3c,d). With pure fully hydrated OPPC, the main transition temperature is found at $-7°C$. The behavior of these ratios has been noted on unsaturated lipids. [14, 15, 16] It is due to the increase in intensity and bandwidth of the $\nu_a(CH_2)$ mode at 2878 cm^{-1} caused by the appearance of a new band at 2881 cm^{-1} between -20 and $-15°C$ where OPPC adopts the L_c^{II} conformation (for $T < -19°C$, the conformation is L_c^I). [17] This conformation seems to favor better packing of the chains of OPPC with respect to DPPC. For temperatures lower than $-20°C$, the cis double bond bending certainly causes an incompatibility between both chains of OPPC which does not permit packing so dense as in the case of DPPC. The difference between R ratios of DPPC and OPPC increases as the temperature decreases which shows that the effect of the bend of the oleoyl chain is more sensitive for temperatures at which the packing could be optimum. The r ratio of OPPC (Fig. 3d) is always slightly higher than that of DPPC in each phases. This may be produced by an increased mobility of the methyl end groups of the oleoyl chains which are much more free than that of the palmitic chain.

These results do not presage with certainty that H–bonds between DPPC and $\alpha--T$ occur in multilayers. Nevertheless, the third component obtained by curve fitting in the C=O region of the FT–IT spectrum after deconvolution can be attributed to vibrations of carbonyl groups of DPPC hydrogen bonded to the hydroxyl group of $\alpha - T$. [17] It is likely that the chain packing of OPPC is different from that of DPPC, as well as the conformation of the glycerol backbone. In particular, the area per molecule is higher for OPPC. It could then be more hydrated in MLVs than DPPC. [18] The phenomenon could be the same for DPPC/$\alpha - -T$ MLVs, as shown by the comparison with OPPC in the $\nu_{C=O}$ region.

2.2.   LIPID - SODIUM LASALOCID

Same experiments have been done for a mixture of DPPC/sodium lasalocid (Las-Na) [19] at a 10:1 molar ratio in water at pH 6. With the addition of Las-Na the pre-transition of the DPPC is not observed and the main transition temperature T_m is reduced by 4°C. In the temperature range below the main transition (10°C - 40°C), the addition of Las–Na causes a significant decrease in the intensity ratio. At temperatures above the main transition, this ratio is significantly smaller than the ratio obtained with DPPC in water. Addition of Las–Na in the DPPC multilayers increases also slightly the proportion of gauche bonds for temperatures below the order — disorder transition temperature as it has been observed in the C-C region. Above the main transition, the intensity ratios are very similar to those of pure DPPC. The antibiotic strongly perturbs the DPPC bilayer with an increase of the fluidity above and below the main transition.

With addition of Las–Na no change is observed in the interfacial region of the MLVs by FT–IR. But, very small changes are observed in the polar region: near 1069 cm^{-1} (R-O-P-R' group) and near 1180 cm^{-1} (C-C(=O)-O-C group). Intensity of both vibrations is concentration dependent, it decreases as the concentration of antibiotic increases. These results suggest that Las-Na molecules are inside the hydrophobic core of the membrane and interact with DPPC molecules.

3.   WATER IN BIOMEMBRANES

It is known that in the gel phase $L_{\beta'}$, the hydration of DPPC molecules is maximal at around 18 H_2O/DPPC [21, 22], but another value has also been found : 11 H_2O/DPPC. [23] The threshold for maximum hydration corresponds to the swelling limit of the membranes. For water concentrations higher than this limit, for which excess water (bulk water or aqueous phase) is present, the interstitial layer thickness (d) of DPPC as well as its main phase transition temperature are constant (1.7 nm and 41°C, respectively). When the water concentration is smaller than that corresponding to the swelling limit, d becomes smaller and the system

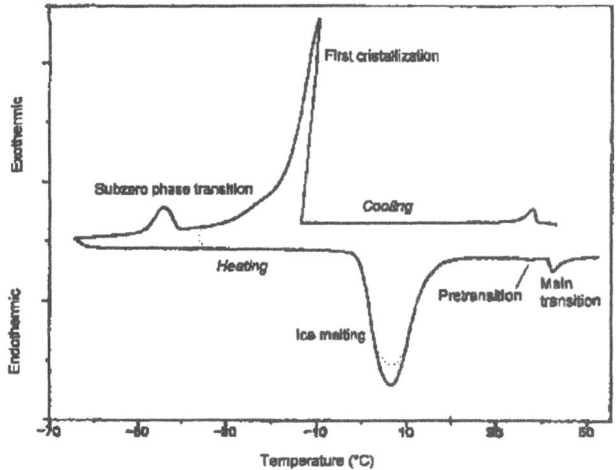

Figure 4. DSC thermogram betwen +50°C and -60°C for an aqueous solution of DPPC at 3°C/min. The dotted line is obtained when the sample is cooled only at -30°C.

does not contain excess water. The aqueous phase is believed to have the properties of pure water, whereas the characteristics of interstitial water are specific. In fact, the hydrogen–bond networks of polar molecules with interstitial water and bulk water, respectively, are different. [24, 25, 26, 27, 28]

Figure 4 represents a complete cycle between $+50°C$ and $-60°C$ obtained by differential calorimetry (DSC) for an aqueous dispersion of DPPC. All the phase transitions are evidenced: the main water crystallization near $-15°C$, the sub-zero phase transition at $-40°C$, the melting of ice at $0°C$, as well as the pre–transition and the main transition of DPPC at 35 and $41°C$, respectively. The thermogram leads one to conclude that, if the pre–transition and main transitions are not affected by this cyclic thermal treatment, the system must be in the gel phase $L_{\beta'}$ just after melting of ice. Consequently, this indicates, that the reverse of the sub–zero phase transition is necessarily hidden by the melting of ice on heating. If the sample is cooled to $-30°C$, melting of ice occurs at the same temperature as above, but with a smaller enthalpy as we can see with the dotted line on fig. 4. The difference has nearly the same value as the sub–zero enthalpy. The fact that the reversible sub–zero phase transition occurs at $-40°C$, and that the melting peak of ice, at $0°C$, is not preceded by an endothermic sub-zero transition peak on heating, leads one to conclude that transition at $-40°C$, is mainly concerned with

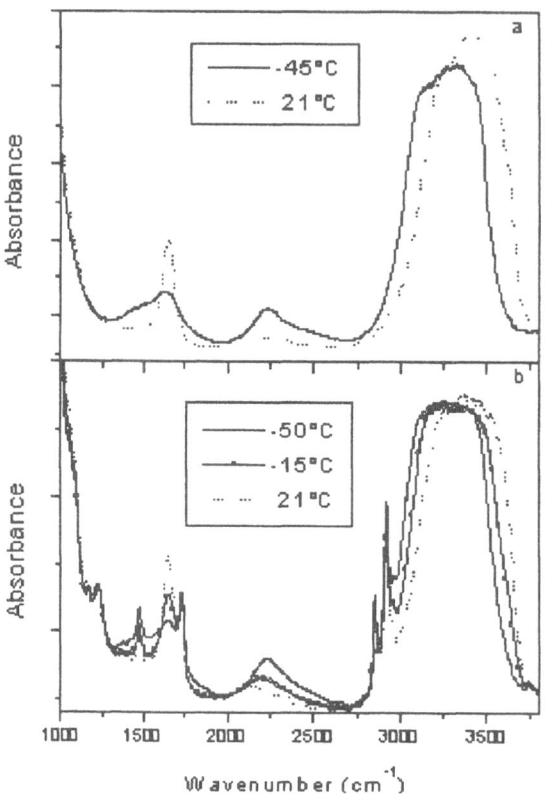

Figure 5.   FT–IR spectra of: a) pure water at 21 and 45°C and b) multilamellar DPPC at 21, -15 and -50°C.

crystallization of water. In figure 5a, we present infrared spectra of pure water recorded at 21°C and -45°C, i.e. in the liquid and in the solid phase, respectively. Figure 5b shows infrared spectra of DPPC MLVs between 1000 and 4000 cm^{-1}. The different phases of the system are represented by spectra recorded at room temperature (21°C), at -15°C (i.e. between the first crystallization and the sub-zero phase transition), and at -50°C (i.e. below the sub-zero phase transition). The bands of water are due to the O-H stretching vibrations ν_{O-H} (3370 cm^{-1}), the O-H bending vibration δ_{O-H} (near 1644 cm^{-1}) and the combination mode $\nu_L + \delta_{O-H}$ (around 2150 cm^{-1}), where ν_L is the so–called libration mode. At room temperature, frequencies, bandwidth at half height and relative intensity of bands of water in MLVs in the gel phase $L_{\beta'}$ are similar to those of pure water, indicating that the most part of water is in the liquid state at this temperature. [29] Both figures show that crystallization of water is marked by sharp discontinuities: the frequency of the combination mode $\nu_L + \delta_{O-H}$ increases abruptly from 2140 to 2200 cm^{-1}, and the intensity of the O-H deformation mode (1644 cm^{-1}) strongly decreases. At the first crystallization, the O-H stretching frequency decreases from 3370 to 3290 cm^{-1}, similarly the shift observed at the crystallization of pure water.

At -50°C, similar changes as those observed at the first crystallization, but with a smaller extent are observed: the $\nu_L + \delta_{O-H}$ frequency is shifted up to 2234 cm^{-1} and the intensity of the δ_{O-H} decreases again, while the ν_{O-H} frequency is shifted down to 3270 cm^{-1}. All the frequencies of water bands in MLVs at -50°C (and up to -60°C) are very closed to the corresponding frequencies of pure ice [29], suggesting that the most part of water in MLVs is crystallized when cooled down to -50°C. This implies that a part of water remains in the liquid state between the first crystallization and the sub-zero phase transition. Therefore, although the presence of lipids is necessary for the sub–zero transition to occur, they are not directly involved in the sub-zero phase transition. This finding is corroborated by the absence of discontinuity of the lipid bands when transition occurs. On heating from -60°C to room temperature, no change occurs up to 0°C, the fusion of water, where the frequencies and intensities return to their room temperature values. These FT–IR results gave direct evidence that the sub–zero phase transition involves water, essentially.

As no modification altered the IR vibrations of DPPC ($\nu_{C=O}$, $\nu_{PO_2^-}$, ν_{CH_2}) at -15 and -40°C, one could neglect the interactions between lipids and water. With this assumption and the value of enthalpy of melting of ice (5.98 kJ/mol) it was then possible to evaluate the number of water molecules involved in the sub–zero phase transition. The mean value in the range from the 27 H_2O/DPPC molar ratio up to pure water is around 9.

4. CONCLUSION

The experiments described above show that Raman scattering and FT-IR absorption (two non destructive techniques), sometimes used with additional techniques as DSC, are powerful tools to analyze interactions between host molecules or water and lipids and seen the effect on the thermotropic behavior of lipids. We also analyzed the effect of confinement of water into MLVs and the peculiar properties this late in these conditions.

References

[1] S.J. Singer and G.L. Nicholson, *Science* **175** (1972) 720.

[2] C. Tanford, in: *The Hydrophobic Effect* (John Wiley and Sons, New York, 1973).

[3] C. Yeagle, in: *The Structure of Biological Membranes*, (CRC Press, Boca Raton, 1992).

[4] G. Turrell, in: *Infrared and Raman Spectra of Crystals*, (Academic Press, New York, 1972).

[5] B.P. Gaber and W.L. Petitcolas, *Biochim. Biophys. Acta* **465** (1977) 260.

[6] N. Yellin and I.W. Levin, *Biochim. Biophys. Acta* **489** (1977) 177.

[7] E. Mushayakarara and I.W. Levin, *J. Phys. Chem* **86** (1982) 2324.

[8] A. Blume, W. Hübner and G. Messner, *Biochemistry* **27** (1988) 8239.

[9] E. Mushayakarara and I.W. Levin, *J. Phys. Chem* **86** (1982) 2324.

[10] J.L. Lippert and W.L. Petitcolas, *Proc. Natl. Acad. Sci.* (USA) **68** (1971) 1572.

[11] L.J. Lis, J.W. Kauffman and D.F. Shriver, *Biochim. Biophys. Acta* **406** (1975) 453.

[12] D.F.H. Wallach, S.P. Verma and J. Fookson, *Biochim. Biophys. Acta* **559** (1979) 153.

[13] T. Lefèvre and M. Picquart, *Biospectroscopy* **2** (1996) 391.

[14] A. Seelig and J. Seelig, *Biochemistry* **16** (1978) 45.

[15] M. Suzuki and T. Ozaki, *J. Am. Chem. Soc.* **62** (1985) 1600.

[16] M. Kobayashi, F. Kaneko, K. Sato and M. Suzuki, *J. Phys. Chem.* **90** (1986) 6371.

[17] M. Picquart and T. Lefèvre, *Chem. Phys. Lipids* **95** (1998) 79.

[18] C.B. Berde, H.C. Andersen and B.S. Hudson, *Biochemistry* **19** (1980) 4279.

[19] M. Picquart, *Rev. Mex. Fis.* **46** (2000) 166.

[20] H. Hauser, I. Pascher, R.H. pearson, and S.Sundell, *Biochim. Biophys. Acta* **650** (1981) 21.

[21] M.J. Janiak, D.M. Small and G.G. Shipley, *J. Biophys. Chem.* **254** (1979) 6068.

[22] M.J. Ruocco and G.G. Shipley, *Biochim. Biophys. Acta* **691** (1982) 309.

[23] C. Grabielle–Madelmont and R. Perron, *J. Colloid and Interface Sci.* **95** (1983) 471.

[24] J.N. Israelachvili, in: *Intermolecular and Surfaces Forces* (Academic Press, London, 1991).

[25] G.E. Walrafen , in: *Water, a comprehensive treatise*, ed. F. Franks, (Plenum Press, New York, 1973) Vol. 1 chapter 5.

[26] M. Lafleur, M. Pigeon, M. Pézolet and J.P. Caillé, *J. Phys. Chem.* **93** (1989) 1522.

[27] S. Kint, P.H. Wermer and J. Scherer, *J. Phys. Chem.* **96** (1992) 446.

[28] J. Grdadolnik, J. Kidricand and D. Hadzi, *J. Mol. Struct.* **93** (1994) 322.

[29] J.M. Baker, J.C. Dore, J.M. Seddon and A.K. Soper, *Chem. Phys Lett.* **256** (1996) 649.

POLYMER CHAIN COLLAPSE IN SUPERCRITICAL FLUIDS

C. Ortiz–Estrada
Chemical Engineering Department, Instituto Tecnologico de Celaya, Celaya
ESQIE, Instituto Politécnico Nacional and Universidad Iberoamericana A. C., México
D. F. MEXICO Guanajuato, 38010 México

G. Luna–Bárcenas
Laboratorio de Investigación en Materiales, CINVESTAV Unidad Querétaro, Querétaro, 76230 México
E-mail:gluna@arcos.qro.cinvestav.mx

G. Ramirez–Santiago
Instituto de Física, UNAM, PO Box 20-364, México 01000 D. F., México
Chemical Engineering Department, The University of Texas at Austin, Austin TX 78712 U.S.A.

I. C. Sanchez
Chemical Engineering Department, The University of Texas at Austin, Austin TX 78712 U.S.A.
sanchez@che.utexas.edu

J. F. J. Alvarado
Chemical Engineering Department, Instituto Tecnologico de Celaya, Celaya Guanajuato, 38010 México

Abstract   Recent computer simulations of a polymer chain in a solvent have provided evidence, for the first time, of polymer chain collapse near the lower critical solution temperature (LCST). Motivated by these results, we have studied further this system to understand the effect of solvent and monomer sizes, chain length, and solvent and monomer energetic interactions. By means of extensive Monte Carlo simulations, the mean

Developments in Mathematical and Experimental Physics,
Volume B: Statistical Physics and Beyond
Edited by Macias *et al.*, Kluwer Academic/Plenum Publishers, 2003

radius of gyration R_g and end–to–end distance R, are calculated for a single chain in a solvent over a broad range of volume fractions, pressures and temperatures. Our results indicate that in general, the chain collapses as temperature increases at constant pressure, or as density decreases at constant temperature. A minimum in R_g and R occurs near the LCST and slightly above the coil–to–globule transition temperature (C-GTT), where the chain adopts a quasi–ideal conformation, defined by the balance of binary attractive and repulsive interactions. At temperatures well above the LCST, the chain expands again suggesting an upper critical solution temperature (UCST) phase boundary above the LCST forming a closed-immiscibility loop. However, this observation strongly depends on the solvent–to–monomer size ratio.

Keywords: Polymers, chain collapse, supercritical fluids, critical solution temperature, polymer–solution phase diagram, simulation of polymers.

1. INTRODUCTION

As it is well known, polymer solutions also exhibit lower critical solution temperature (LCST) or thermally induced phase separation when temperature is raised. LCSTs have been observed in strongly interacting polar mixtures, for example aqueous solutions, as well as in weakly interacting non–polar polymer solutions. An LCST may also be produced by the mixture's finite *compressibility*. Unlike the upper critical solution temperature (UCST), which is driven by unfavorable energetics, a thermodynamics analysis shows that LCST is an entropically driven phase separation.

LCSTs usually occur in the vicinity of the vapor–liquid critical temperature of the pure solvent. Specifically, they are typically observed at about $0.7T_c^*$ to $0.9T_c^*$. Of particular relevance is the phase behavior of polymer solutions in supercritical fluids (SCFs), since most of these systems exhibit an LCST.

The development of a fundamental understanding of the phase behavior of polymer solutions in SCFs is a theoretical challenge of great practical interest[1]. The solution behavior is complex due to large values of free volume, isothermal compressibility, volume expansivity and concentration fluctuations. Practical SCF applications involving LCST phase behavior include polymer fractionation, impregnation and purification, polymer extrusion and foaming, formation of materials by rapid expansion from supercritical solution and precipitation with a compressed fluid antisolvent, dispersion as well as emulsion polymerization, and formation of emulsions and microemulsions [2,3].

It was conjectured some time ago that chain collapse should also be observed near an LCST in an analogous way as it occurs near an UCST [4]. The implications of this polymer physics problem are of great relevance to SCF technology. In 1997, Luna–Bárcenas, *et. al.* [5], reported for the first time, from extensive numerical calculations, evidence of polymer chain collapse near the LCST upon heating the polymer solution at constant pressure. By investigating single chain architecture, the phase behavior of more concentrated polymer solutions was predicted. In other words, the physics of chain collapse near LCST captures the macroscopic phase separation behavior in a finite concentration polymer solution. Luna–Bárcenas, *et. al.* observed that the collapse of a single polymer chain correlates well with *coil–globule transition temperature (CGT-T)* and an occurrence of an LCST phase boundary. It was also observed that upon further heating the collapsed chain expanded again suggesting the presence of a closed immiscibility loop. Later, this closed loop was corroborated by direct phase separation simulations using an expanded Gibbs ensemble formalism [5]. However, the above studies were restricted to a fixed chain length, energetic interactions and solvent to monomer size ratio. Motivated by these findings, in this work we report some results of more extensive numerical simulations by investigating the effect, that energetic interactions and monomer to solvent size, have on the phase diagram of a mixture.

2. MOLECULAR MODEL

The system studied in this work consists of a single freely jointed chain immersed in a solvent medium and is analogous to the infinite dilute regime of a polymer system. This regime consists of a polymer system in which chains are far from each other to avoid interchain interactions, that is, the chains act as individual entities. To simulate the intermolecular interactions we used the typical Lennard–Jones (LJ) potential that is defined as,

$$U_{i,j}(r) = \begin{array}{ll} 4\epsilon_{ij}\left[\left(\frac{\sigma_{ij}}{r}\right)^{12} - \left(\frac{\sigma_{ij}}{r}\right)^6\right] + 0.01632\epsilon_{ij}, & \text{if } r \leq 2.5\sigma_{ij}. \\ 0 & \text{otherwise.} \end{array}$$

In this equation r is the distance between molecules, σ_{ij} and ϵ_{ij} represent the parameters of the potential. Since we have introduced a cutoff of 2.5σ in the interaction LJ potential, it is equal to zero for larger distances. This is equivalent to an upwards shift in the entire potential . The phase and critical behavior of this LJ model have been studied by Smit [6] who reported a reduced critical temperature $T^* = (K_B T_c/\epsilon) = 1.08$, a reduced critical density $\rho^* = \sigma^3 \rho_c = 0.31$, and a reduced critical pressure $P^* = P_c \sigma^3/\epsilon = 0.10$. In this paper, we report results corresponding to a constant chain length of $N = 20$ segments or monomers. The strength of the energetic interactions has been varied by changing systematically the ratio $\epsilon_{11}/\epsilon_{22}$. Here, the subscript 11 refers to the unbounded monomer–monomer interaction while the subscript 22 is related to the solvent–solvent interaction. In the same manner, the monomer–solvent size effect is considered by varying the ratio σ_{11}/σ_{22}, with σ_{11} the monomer size and σ_{22} the solvent size. The way the site density is usually defined, ($\rho^* = \sigma^3 \rho$) is not appropriate when dealing with objects of dissimilar sizes. Instead, we use the volume fraction, $\eta = (site\ volume)/(total\ volume)$ as a more natural variable since it takes into account the volume of the objects, whereas the site density does not.

To study the polymer chain collapse we used the continuum configurational bias (CCB) Monte Carlo algorithm. This method consists in cutting the chain at a random site. A portion of the chain is then deleted from this site to one of the ends of the chain. Finally, the chain is regrown site by site until its original length is restored. A more detailed description and explanation of this algorithm is presented in reference [7].

3.   RESULTS AND DISCUSSION

3.1.   SOLVENT–TO MONOMER SIZE RATIO EFFECT:(SMALL SOLVENT–BIG MONOMER)

To understand better the effect of the monomer–to–solvent size ratio on the mixture's phase behavior, we considered a monomer segment volume that is twice the solvent volume. This is equivalent to consider a ratio $\sigma_{11}/\sigma_{22} = 1.26$, The energetic interactions were chosen such that the interactions between monomer-monomer and solvent–solvent are equal, that is, $\epsilon_{11}/\epsilon_{22} = 1$ keeping this ratio constant.

Figure 1 shows the mean square end–to–end distance $< R^2 >$ of the chain as a function of the system volume fraction at several reduced temperatures, that were chosen in the vicinity of the pure solvent critical temperature $T_c^* = 1.08$. For reasons of comparison it is important to note that in figure 1 the data labelled with the legend "BASE" represent the results for a symmetric mixture, that is, $\epsilon_{11}/\epsilon_{22} = 1$ and $\sigma_{11}/\sigma_{22} = 1$, as reported in reference [5]. At high densities, the chain adopts a coil-like conformation that approaches the athermal or infinite temperature limit. The chain collapse at low densities suggest that solvent quality diminishes as the solvent density or volume fraction decreases. This behavior is experimentally observed in a pure supercritical fluid. For instance, the square of the solubility parameter (cohesive energy density per unit volume) decreases when the density also decreases. It is interesting to note that going from low to high volume fraction the *small*

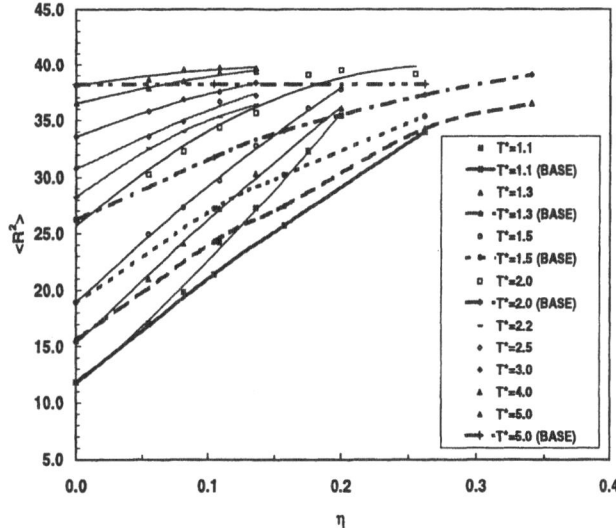

Figure 1. Mean square end–to–end distance versus volume fraction for a small solvent–big polymer system. See text for a discussion of this figure.

solvent–big polymer mixture expands more rapidly than the symmetric mixture at fixed temperature. This suggest that a smaller solvent with similar energetic interactions compared to a bigger one acts as a *better solvent* since the chain is more solvated. Also note from Figure 1 that the increase in chain dimensions with temperature at constant volume fraction — pressure must also increase to maintain constant the density — is consistent with the idea that attractive energetics become less important at high temperatures, that is, $\epsilon/(k_B T) \to 0$. When the chain collapses to enhance favorable intrachain attractive interactions, it does so at the expense of losing chain conformational entropy. Chain connectivity brings chain segments into close proximity to one another — the so called correlation hole — enhancing the effects of intrachain forces relative to chain–solvent interactions. The remarkable behavior of chain collapse with temperature is shown in figure 2.

At constant pressure chain dimensions go through a minimum suggesting that a phase boundary is being approached. It has been shown recently [5] that this minimum represents the LCST of the mixture. However, in our case *small solvent–big polymer*, the chain's minimum dimension is bigger than the symmetric case, that is, *same size solvent and monomer*, which is in agreement with the observation made above

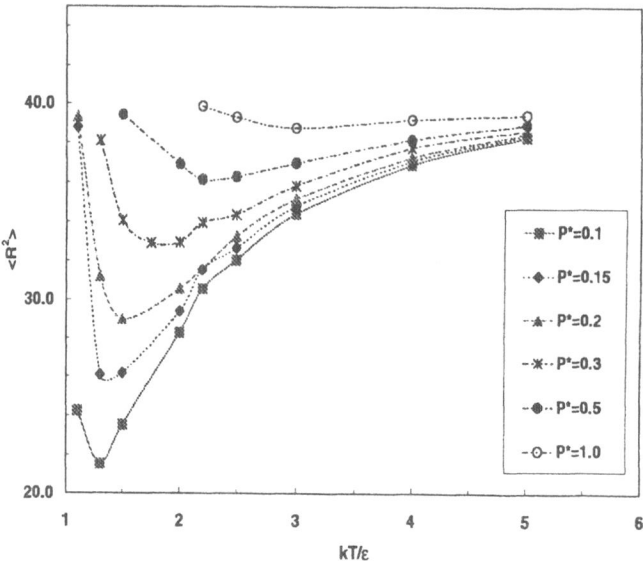

Figure 2.   Mean square end–to–end distance versus temperature for a small solvent–big polymer system. These results are discussed in the text.

from figure 1, that smaller solvent solvates better the chain. The expansion of the chain upon further heating — see figure 2 — suggest the existence of a one–phase region. These observations were already pointed out in reference [5]. In fact they demonstrated, by direct simulation of phase equilibria, the existence of a closed immiscibility loop in the polymer solvent phase diagram.

3.2.   SOLVENT–TO–MONOMER SIZE RATIO EFFECT (BIG SOLVENT SMALL POLYMER CASE)

In contrast to the previous subsection, we now consider the effect of a solvent that has twice the volume of the monomer unit in the chain, that is $\sigma_{11}/\sigma_{22} = 0.794$. In figure 3 we show the behavior of the mean square end–to–end distance as a function of temperature. It is worth noting that at a given temperature and pressure, but at temperatures near the solvent's critical point, the chain collapses when compared to the *small solvent–big polymer case*. This fact would imply that the mixture would be phase separated until high enough temperatures are reached at which the mixture will again be miscible. The predicted phase behavior is shown in figure 4 on a P–T plane. Figure 4 also shows the effect of

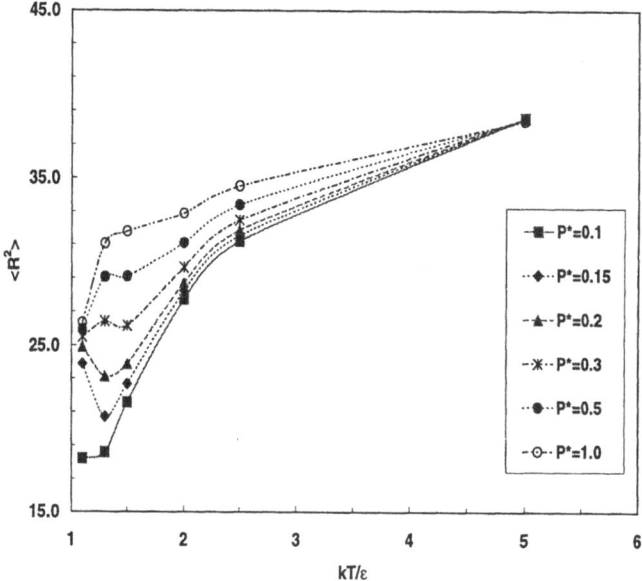

Figure 3. Mean square end–to–end distance versus temperature for a big solvent–small polymer system. See text for an interpretation of this figure.

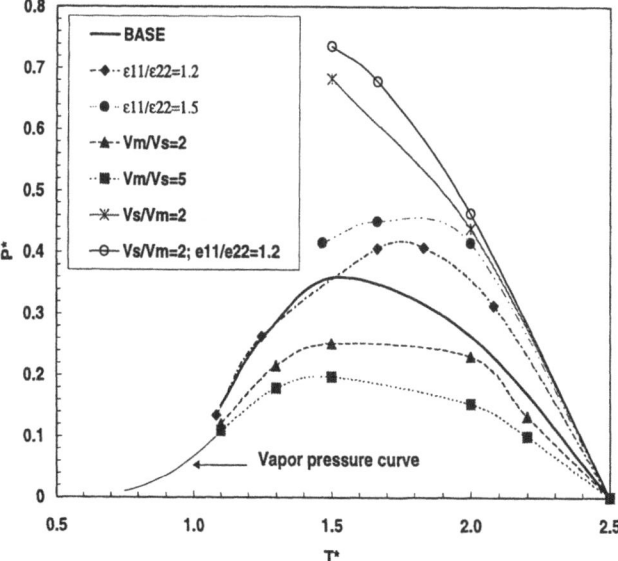

Figure 4. Predicted pressure versus temperature phase diagram.

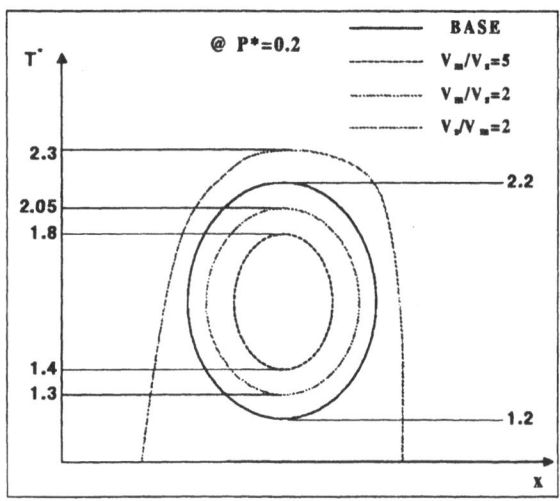

Figure 5. Expected temperature versus composition diagram. The reduced pressure of the system is $P^* = 0.2$.

different energetic interactions. Two cases are considered: $\epsilon_{11}/\epsilon_{22} = 1.2$ and $\epsilon_{11}/\epsilon_{22} = 1.5$. In both cases, the non-bonded monomer-monomer energetic interactions are stronger than the solvent-solvent and monomer-solvent interactions. These conditions mimic a *worse* solvent case with respect to the symmetric mixture. Finally, a few words on the polymer solution phase diagram. The collapsing chain — emulating the infinite dilution — signals the occurrence of an LCST phase boundary in a more concentrated (finite) solution. Further heating favors the expanding of the chain, which also indicates the existence of a nonhomogeneous region. The macroscopic picture of the above structural changes — collapsing-expanding — of the chain results in a closed immiscibility loop of varying size and shape. For instance, the *small solvent–big polymer* case predicts a smaller inmiscibility region whereas the *big solvent–small polymer* case predicts a bigger immiscibility window. For illustrative purposes, the *expected phase diagram* for the different cases explained above is depicted in figure 5.

Acknowledgments

Financial support of this work has been provided by CONACyT, Grants J33262 (G. Luna-Bárcenas) and 25298–E and G32723–E (G.

Ramírez–Santiago). We also received financial support from JIRA–CIN-VESTAV and the NSF Science & Technology Center for the Utilization of Carbon Dioxide in Manufacturing at the University of Texas at Austin.

References

[1] Folie B. and Radosz M. *Ind. Eng. Chem. Res.* **34** (1995) 1501.

[2] McHug M. A. and Krukonis V. J. *Supercritical Fluids Extraction* (Butterworth-Heinemann, 1994).

[3] Prausnitz J. M., Lichtenthaler R. N. and Gomes de Azevedo E., *Molecular thermodynamics of fluid phase equilibria* (Pretince Hall, 1999).

[4] Sanchez I. C. *Macromolecules* **12** (1979) 980.

[5] Luna–Bárcenas G., Meredith J. C., Gromov D., de Pablo J. J., Johnston K. and Sanchez I. C., Jour. Chem. Phys. **107** (1997) 10782.

[6] Smit B. Jour. Chem. Phys. **96** (1992) 86392.

[7] de Pablo J. J., Laso M. and Suter U. J., Jour. Chem. Phys. **96**(1992) 2395.

III

GRANULAR MATERIALS AND QUANTUM SYSTEMS

AMPERIAN MAGNETISM IN THE DYNAMIC RESPONSE OF GRANULAR MATERIALS

Rubén G. Barrera

Instituto de Física,
Universidad Nacional Autónoma de México,
Apartado Postal 20–364, 01000 México D. F. México.

Augusto Garcia–Valenzuela

Centro de Instrumentos,
Universidad Nacional Autónoma de México,
Apartado Postal 70–18, 04510 México D. F. México

Abstract   We study the coherent reflectance of electromagnetic waves from a random system of identical spheres with radius comparable to the wavelength of the incident radiation. An effective–medium theory for this system is developed and it is found that the effective–medium must posses an effective magnetic permeability, even if the spheres are non-magnetic, in order to be consistent with continuum electrodynamics. The physical origin of this magnetic effect is discussed and we conclude that it is due to the induction of closed currents in the spheres, being then analogous to the mechanism proposed by Ampère when he tried to explain the origin of magnetism. It turns out that the effective magnetic permeability as well as the effective electric permittivity depend on the angle of incidence and the polarization of the incident wave. We derive formulas for the coherent reflectance from a half–space and display numerical results.

Keywords: granular matter, electrodynamics, optical properties, random system, effective-medium theory

1.   INTRODUCTION

The concept of an effective medium has been extremely useful in the description of the electromagnetic response of granular matter [1]. By granular matter we will understand a very general class of materials

Developments in Mathematical and Experimental Physics,
Volume B: Statistical Physics and Beyond
Edited by Macias *et al.*, Kluwer Academic/Plenum Publishers, 2003

147

composed by granular inclusions of one type of material embedded in an otherwise homogeneous matrix of another type of material. By electromagnetic response we mean the polarization and magnetization processes induced in the material by an externally applied electromagnetic field. In electrodynamics of continuous media one introduces the concept of polarization and magnetization fields. They are called material fields because they are defined only in the regions occupied by the materials, they are attached to their presence, and outside these regions these fields vanish. The optical properties of the material are determined by the manner in which these material fields respond to the applied one, and this response is usually given in terms of response functions like the dielectric function and the magnetic susceptibility, which will be generally referred as the optical coefficients of the material.

Furthermore, one might think that behind all these concepts there is the assumption of a continuous material and that the values acquired by the material fields correspond to a quantitative measure of the induced polarization and magnetization phenomena. Nevertheless, one also knows that in essence all matter is granular, after all, one can think that any piece of matter is made of a large collection of small grains, these grains being the atoms and molecules. Therefore, due to this granular structure the induced electromagnetic field within any material is a highly varying function of space and time. This field is usually called the *microscopic* field. But when this microscopic field is decomposed as the sum of an average component plus a fluctuation component, one finds that the length scales of their spatial variations are very different, while the average component varies on a length scale of the order of the wavelength of the incident field, the fluctuation component varies on a length scale of atomic dimensions. Now, the fields that appear in Maxwell's equations of continuous media correspond only to the average component, the fluctuation component is neglected. The average component is also called the macroscopic field and one refers to the equations that govern its behavior as the macroscopic Maxwell's equations. As a consequence, all the laws derived in electrodynamics of continuous media, like Snell's law, Fresnel's relations and Poynting's theorem, neglect the contribution due to the field fluctuations, providing only relations between the average (macroscopic) values of the electromagnetic and material fields. Thus the concept of continuity in the macroscopic equations is somewhat artificial and is the result of an averaging procedure that smooths out the space–time variations of the fields by neglecting the field fluctuations. But in ordinary materials the power carried along by the average component is much greater than the one carried along by the fluctuations, and this fact is what justifies the successful application of continuous electrodynamics to ordinary materials. As a consequence, one can conclude that the concept of continuity in macroscopic electrodynamics is a matter of scale based in the actual possibility of neglecting the contribution of the small fluctuations of the fields caused by the molecular "granularity" of matter. Nevertheless, this does not mean that the field fluctuations are undetectable, because if this were true we would not be able to see, for example, a blue sky.

Now we move to the problem of the optical properties of granular materials where the characteristic size of the individual grains is not longer of atomic dimensions but is rather of macroscopic dimensions, and their optical properties are described by macroscopic electrodynamics. First we will assume that although the characteristic size of the grains is macroscopic, it is still much smaller than the wavelength of the incident radiation. In this case the power carried along by the average component of the electromagnetic field is still large as compared to the one carried along by the fluctuations; although not as large as in the case of ordinary materials with "molecular" granularity. If we now concentrate our attention in the physical description of *only* the average component of the fields, we can ask ourselves if it is now possible to extend the concept of the continuity of matter and to define a continuous medium in which the average component of the field behaves exactly in the same manner as in the actual granular material. This artificial continuous medium is commonly known as *effective* medium, in which effective material fields can be defined and whose response to an applied external field yield, for example, an effective dielectric function and an effective magnetic susceptibility. From this perspective one could regard the material fields of ordinary materials in macroscopic electrodynamics, also as properties of an artificial effective continuous medium which describes correctly the behavior that the average field has in the actual material with "molecular granularity". The main advantage of an effective–medium approach is that one can immediately use all the results of continuous electrodynamics by simply setting, in the relevant expressions, instead of the macroscopic response functions the effective response functions, and to certain extent one forgets about the granularity of the material. One has only to be careful about certain aspects of the physical interpretation of the results. For example, in macroscopic electrodynamics one interprets the imaginary part of the response functions as a quantity that is proportional to the absorption of energy by the system, in the case of an effective medium of a granular material, the imaginary part of the effective response functions is proportional not only to the energy that is absorbed but also to energy that is scattered. In this way, the energy balance forces one to look at the energy flux carried by the fluctuations, as an energy flux that is "taken away" from the flux carried by the average component of the field.

The problem is now to find a relationship between the effective response functions and the actual geometric and optical parameters of the grains and the matrix, as well as the statistical parameters that describe the way in which grains are mixed into the matrix. This is the problem that has attracted the attention of many researchers for more than a century, and has required the efforts of theoretical and experimental physicists as well as applied mathematicians and engineers [2]. Besides its interest as a problem in basic physics, its solution can be used in a wide variety of applications. First, because its range of application extends beyond the field of optical properties and comprises all physical properties involving a linear response to an external field, like in the elastic, electric, thermal, or hydrodynamic properties of either granular composites, rocks, emulsions, suspensions or colloids. The interest lies then in the calculation, in this type of systems, of properties like: the effective stress-strain tensor, the effective electrical conductivity, the effective thermal conductivity, the effective chirality or the effective viscosity. Second, because the knowledge of the relationship between the effective response properties and the parameters that characterize a granular system, opens the possibility for the construction and design of novel type of materials fulfilling requirements and specifications that cannot be found in ordinary materials. Although there has been a significant progress in the solution of this problem and new type of materials with unexpected properties have been produced and designed, we are still far from claiming that the problem has been finally solved. There is a wide collection of expressions, called "mixing rules", that propose explicit expressions that relate the parameters characterizing a granular system and its effective response, nevertheless their range and conditions of validity as well as the microstructure that is assumed for their derivations, are issues that very often are not clear. Also, the experimental characterization of the microstructure of a granular system is not an easy task, and one usually ends up with only a few from the total set of relevant parameters.

Finally, we address essentially the same problem of extending the idea of a continuous effective medium for the description of the optical properties of a granular system, but now when the characteristic size of the grains is of the same order of magnitude as the wavelength of the incident radiation. In this case the power carried along by the fluctuations of the field might be as large as the one carried along by the average component, and the average field is now called the coherent field, while the field fluctuations are called the diffuse field. Nevertheless, one can still ask oneself if in this case the behavior of the average component of the field can still be described by a continuous effective medium. There are several contributions towards the solution of this more complicated problem in which the scattering processes play a more important role. One finds in the literature explicit expressions that relate, for example, the effective index of refraction of a dilute system of randomly located identical spheres with the forward scattering amplitude of an individual sphere. Probably, the most popular derivation of this relation is the one given by van de Hulst in his book [3]. There have been also efforts to generalize this relation to systems with a larger concentration of spheres [4] or to different geometries, like a spherical matrix with spherical inclusions [5].

However, there are critical remarks about the use of the effective index of refraction derived by van de Hulst in expressions like the Fresnel's relations that yield the reflection and transmission amplitudes from a slab in terms of the index of refraction and the optical coefficients of the material. For example, C. Bohren has considered the simple case of a plane wave at normal incidence into a slab containing a dilute concentration of randomly located identical polarizable spherical inclusions, and he has calculated the coherent component of the reflected and transmitted fields [6]. He finds that in order to calculate the amplitudes of these fields by replacing the slab by a continuous medium with an effective index of refraction and Fresnel's relations, it would be necessary to define two different index of refraction: one for reflection and one for transmission. But instead of doing that he proposes to choose another two different optical coefficients: an effective electric permittivity and an effective magnetic permeability. This last one is proportional to the difference between the forward and backward scattering amplitude of an individual sphere. The problem is to justify how come a composite made of two nonmagnetic components turns out to be magnetic. The main objective of this paper is to clarify this issue as well as to extend the calculation to non–normal incidence. We do this by using scattering–wave and Mie theories to calculate the reflected and transmitted fields from a thin slab with containing a dilute concentration of randomly located identical polarizable spheres, and then we identify the current distributions that may act as sources of these fields. We find that these sources should be given by a superposition of open and closed currents induced in the spheres. The closed currents are induced by the time variations of the magnetic field, and in this sense they represent a true *bona fide* magnetic response of the system. We find that both the effective electrical permittivity and the effective magnetic permeability depend on the angle of incidence and the polarization of the incident field. Therefore they cannot be regarded as intrinsic properties of the granular system, nevertheless they provide the basis for the calculation of the reflected and transmitted fields. Also, their product being proportional to the square of the effective index of refraction turns out to be independent of the angle of incidence and the polarization of the incident beam, and its expression in terms of the scattering properties of the individual spheres coincides with the one derived by van de Hulst.

One might call this type of magnetic response in a granular system: *Amperian magnetism*, because in the beginning of electrodynamics, A. M. Ampère had envisioned the physical origin of magnetism as the result of closed currents induced in the molecules by the time variations of the magnetic field. Later on it was found that magnetism was a more complicated phenomenon related more to the spin of the electrons than to the induction of closed currents in the molecules. But in a system like a granular composite one has small macroscopic spheres instead of molecules and the induction of closed currents is more favorable in large spheres than in small spheres. In a more technical language, the contribution of the spheres to the effective magnetic permeability turns out to be proportional to the asymmetry in the scattering amplitude between the forward and the specular direction in the individual spheres, and from Mie theory one sees that this asymmetry increases with the size of the spheres. In the limit of very small spheres the scattering becomes quite isotropic thus the contribution of the spheres to the effective permeability vanishes, and one recovers the non-magnetic character of the effective medium.

A more transparent example of Amperian magnetism is perhaps the recently developed microstructured materials that operate in the microwave region. A particular type of these materials consists of an insulating matrix in which a collection of millimeter–size copper rings are embedded within forming a 3D periodic structure [7]. A time–varying magnetic field induces currents in the rings giving rise to closed currents that are responsible for the magnetic character of the response. This is a beautiful example of Amperian magnetism and shows how a composite of two non–magnetic materials becomes magnetic. By opening the rings with a small gap, there is also an induced capacitance, which together with the inductance due to the induced currents in the rings, gives rise to a resonance phenomenon, and this yields frequency regions in which the effective magnetic permeability becomes negative. Another interesting microstructured material is an insulating matrix in which a collection of very thin long wires are embedded within the matrix forming a well–defined cubic structure. It can be shown that this type of microstructured materials posses an effective dielectric response that is negative for certain frequencies. By combining the wire and the ring structures it has been shown that there are frequency regions in which both the effective electric permittivity and the magnetic permeability are negative yielding a negative effective index of refraction. It has been also argued that in the frequency regions in which the index of refraction is negative, one could use this uncommon property for the construction of a perfect lens.

Finally, we believe that the expressions derived in this paper, although limited to dilute systems, are not a pure and simple curiosity, but on the contrary they may be useful in several applications. For example, there is now interest to follow, in real time, different processes that take place in turbid media, through the changes in their effective index of refraction. Nevertheless, although measurements of the attenuation of light through turbid systems are done routinely in many laboratories, there are few transmission experiments which measure both, the real and imaginary part of their effective index of refraction [8], [9]. A simple and potentially very useful way of measuring the effective index of refraction in turbid media is by critical–angle refractometers [10], [11], [12]. In this method the real and imaginary parts of the effective index of refraction are obtained by inverting the relationship between the reflection amplitude and the effective index of refraction. The naive use of Fresnel expressions to perform this inversion would lead to errors in both, accuracy and interpretation. In this respect, the expressions for the reflection amplitude derived here could be used, together with data of critical–angle refractometers, to obtain not only more accurate results of the optical constants of turbid media, but also to start doing reliable modelling of the correlation between their changes and some of the specific processes that take place within the system.

2.   BASIC CONCEPTS

First we review some basic concepts of linear response in ordinary materials. If the response is linear the polarization $\mathbf{P}$ and the magnetization $\mathbf{M}$ are proportional to the incident field. The relation between $\mathbf{P}$ and $\mathbf{M}$ and the incident electromagnetic field is given, in general, in terms of integral operators with kernels that are non-local in space and time. However, for the case homogeneous and isotropic ordinary materials one can write $\mathbf{P} = \epsilon_0 \chi^E \mathbf{E}$ and $\mathbf{M} = \chi^H \mathbf{H}$, where χ^E and χ^H are scalar algebraic functions. The functions χ^E and χ^H are called the electric and magnetic susceptibilities, respectively, and they are intrinsic properties of the material. The non–locality in space is avoided by relating $\mathbf{P}$ and $\mathbf{M}$ not to the incident fields but rather to the total fields $\mathbf{E}$ and $\mathbf{H}$, which are given by the sum of the incident plus the (average) induced field. The non–locality in time is accounted for by expressing the susceptibilities in the Fourier space of frequencies. Thus for an isotropic and homogeneous material in the presence of an applied field oscillating with frequency ω, the susceptibilities χ^E and χ^H are only functions of ω. The propagation wavevector k of the field within the material is given by $k = (\omega/c)n$, where c is the speed of light and $n = \sqrt{\tilde{\epsilon}\tilde{\mu}}$ is the index of refraction. Here $\tilde{\epsilon} = 1 + \chi^E$ and $\tilde{\mu} = 1 + \chi^H$. The reflection amplitude r is defined as $r = E_r/E_i$ where is the amplitude of the reflected electric field while E_i is the amplitude of the incident electric field. For the case of reflection from a half space the Fresnel's relations are

$$r_{hs}^{TE} = \frac{\tilde{\mu}k_z^i - k_z}{\tilde{\mu}k_z^i + k_z} \tag{1}$$

and

$$r_{hs}^{TM} = \frac{\tilde{\epsilon}k_z^i - k_z}{\tilde{\epsilon}k_z^i + k_z}, \tag{2}$$

where $k_z = k\sqrt{n^2 - \sin^2\theta_i}$, n is the index of refraction of the material, and the superscripts TE and TM denotes transverse–electric and transverse–magnetic polarization, respectively, referring to the cases where the electric or the magnetic field are perpendicular to the plane of incidence.

As mentioned above, in electrodynamics of continuous media the set of Maxwell's equations describe the behavior of only the average component of the electromagnetic field. Moreover, when one tries to extend the idea of continuity to the case of granular composites through the concept of effective medium, one has to properly define the average of the fields. The average procedure smooths out the field variations to a given specified scale. From the experimental point of view this process can be thought as performed by the measuring apparatus when trying to detect highly varying fields. From a mathematical point of view the averaging procedure can be represented by a projection operator acting on the highly varying microscopic field. There are many ways of taking the average of the microscopic field, there is, for example, a spatial average in which a spatial integration of the field times a weight function is performed around any given point in space, there is truncation in Fourier space in which the spatial Fourier transform is truncated up to a maximum cut off wavevector and then is transformed back into real space. In this work we are dealing with a system with randomly located inclusions and we will consider a configurational average, that is, the field at a certain point in space is averaged by "moving around" the location of the spheres. This "moving around" in a system of randomly located spheres generates a finite set of different configurations characterized by a different location of the spheres. The average value is obtained by adding up the values of the field at any given point in space generated by each configuration and then dividing it by the total number of configurations. One should take a sufficiently large number of configurations in order to obtain a stable value for the average. A critical analysis on the dependence of the results on the type of average that is taken is out of the scope of this work, but interested readers can take a look at Ref. [13].

3. FORMALISM

Our approach to the effective medium theory consists of comparing the average scattered fields from a thin slab of the random system of spheres to the radiated fields by an equivalent homogeneous slab when a plane wave is incident on them. By matching the scattered and radiated fields the optical coefficients of the effective medium are obtained. In what follows we briefly describe the main steps in the calculation of the effective optical coefficients. More details can be found in Ref. [14].

First, we consider a dilute random distribution of spherical particles in vacuum (no matrix) contained in a boundless slab region parallel to the XY plane and $-d/2 < z < d/2$. The system is in the presence of an incident plane wave with an electric field given by $\mathbf{E}^i(\mathbf{r}, t) = E_0 \exp i(\mathbf{k}^i \cdot \mathbf{r} - \omega t)\,\widehat{\mathbf{e}}_i$, where $\mathbf{r}$ and t are the position vector and time, respectively, ω is the radial frequency, $\widehat{\mathbf{e}}_i$ is a unit vector in the direction of polarization, $\mathbf{k}^i = k_y^i\widehat{\mathbf{a}}_y + k_z^i\widehat{\mathbf{a}}_z$ is the incident wave vector assumed to lie on the YZ plane, and $\widehat{\mathbf{a}}_x$, $\widehat{\mathbf{a}}_y$, and $\widehat{\mathbf{a}}_z$ are unit vectors along the Cartesian axes of coordinates (see Fig. 1). The electric field satisfies $\widehat{\mathbf{e}}_i \cdot \mathbf{k}^i = 0$, and $|\mathbf{k}^i| = k$, where $k = \omega/c = 2\pi/\lambda$ is the wave number in vacuum, λ is the corresponding wavelength and c is the speed of light. The time dependence $\exp(-i\omega t)$ will be assumed implicit and we use the SI system of units.

The incident field is scattered by the particles, and we assume that their number density is low enough so the independent–scattering approximation is valid. Within this approximation the total scattered field is given by the sum of the fields scattered by each of the particles in the slab region. Therefore, the scattered field $\mathbf{E}^S$ due to a collection of N spherical particles with their centers located at $\{\mathbf{r}_1, \mathbf{r}_2, ..., \mathbf{r}_p, ..., \mathbf{r}_N\}$ can be written as [4],

$$\mathbf{E}^S(\mathbf{r}) = \sum_{p=1}^{N} \int d^3r' \int d^3r'' \,\overline{\overline{G}}_0(\mathbf{r}, \mathbf{r}') \cdot \overline{\overline{T}}\,(\mathbf{r}' - \mathbf{r}_p, \mathbf{r}'' - \mathbf{r}_p) \cdot \mathbf{E}_p^E(\mathbf{r}''), \quad (3)$$

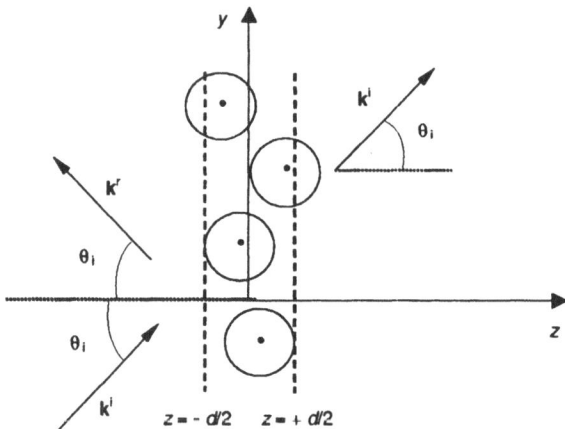

Figure 1.   A slab of a dilute random–system of spheres. The centers of the particles are within the planes $z = -d/2$ and $z = d/2$.

where $\overline{\overline{G}}_0(\mathbf{r}, \mathbf{r}')$ is the dyadic Green's function in free space, $\overline{\overline{T}}(\mathbf{r}', \mathbf{r}'')$ is the transition operator for a sphere [4], and $\mathbf{E}_p^E$ denotes the exciting field. This is defined as the field that drives the scattering process in particle p, that is, the incident field plus the field scattered by the rest of the particles in a region within and around particle p. Thus $\mathbf{E}_p^E$ depends parametrically on the location of the rest $N - 1$ particles.

Since we are assuming a dilute system of particles and thin slab, the exciting field may be approximated as the incident field to the slab, $\mathbf{E}_p^E \simeq \mathbf{E}^i$. By using the plane–wave expansion of the dyadic Green's function and the momentum representation of the transition operator $\overline{\overline{T}}(\mathbf{r}', \mathbf{r}'')$ of an isolated sphere: $\overline{\overline{T}}(\mathbf{p}', \mathbf{p}'')$ and finally averaging the scattered fields with respect to the position of the particles, one arrives to,

$$\langle \mathbf{E}^S(\mathbf{r}) \rangle_{slab} = \begin{cases} \mathbf{E}_+^S \exp(i\mathbf{k}^i \cdot \mathbf{r}) & \text{for } z > d/2 \\ \mathbf{E}_-^S \exp(i\mathbf{k}^r \cdot \mathbf{r}) & \text{for } z < d/2 \end{cases}, \tag{4}$$

where

$$\mathbf{E}_+^S = i\frac{E_0}{2}\rho\frac{(\overline{\overline{1}} - \widehat{\mathbf{k}}^i\widehat{\mathbf{k}}^i)}{k_z^i} \cdot \overline{\overline{T}}(\mathbf{k}^i, \mathbf{k}^i) \cdot \widehat{\mathbf{e}}_i\, d \tag{5}$$

$$\mathbf{E}_-^S = i\frac{E_0}{2}\rho\frac{(\overline{\overline{1}} - \widehat{\mathbf{k}}^r\widehat{\mathbf{k}}^r)}{k_z^i} \cdot \overline{\overline{T}}(\mathbf{k}^r, \mathbf{k}^i) \cdot \widehat{\mathbf{e}}_i\, \frac{\sin k_z^i d}{k_z^i}, \tag{6}$$

where $\mathbf{k}^r = k_x^i \widehat{\mathbf{a}}_x + k_y^i \widehat{\mathbf{a}}_y - k_z^i \widehat{\mathbf{a}}_z$ is the wave vector in the specular direction, $k_z^i = \sqrt{k^2 - (k_x^i)^2 - (k_y^i)^2}$, and ρ is the density of particles. In the averaging procedure we assumed that the positions of the particles are independent of each other (i.e., we ignored the exclusion volume) and that the probability to find a particle with its center inside the volume $d^3\mathbf{r}$ is uniform and given by $d^3\mathbf{r}/V$, where V is the volume of the slab. Eq. (4) means that the scattered field interferes constructively along two directions: $\mathbf{k}^i$ and $\mathbf{k}^r$, independently of the location of the scatterers, for this reason these are the only components of the field that survive after a configurational average.

Equations (5) and (6) can be put in terms of the scattering matrix elements commonly used in describing light scattering from small particles. (The scattering matrix is clearly defined in the book by Bohren and Huffman [15]) First, one recognizes that $(\overline{\overline{\mathbf{1}}} - \widehat{\mathbf{k}}^a \widehat{\mathbf{k}}^a) \cdot \overline{\overline{T}}(\mathbf{k}^a, \mathbf{k}^b) = 4\pi \overline{\overline{F}}(\widehat{\mathbf{k}}^a, \widehat{\mathbf{k}}^b)$, where $\overline{\overline{F}}$ is the far–field scattering dyad, and then express $\overline{\overline{F}}$ in terms of the scattering matrix elements S. When the particles are spherical, there are only two non–zero matrix elements, S_1 and S_2 and the following expression is obtained

$$\mathbf{E}_+^S = -E_0 \gamma \frac{kd}{\cos\theta_i} S(0) \widehat{\mathbf{e}}_i \qquad (7)$$

$$\mathbf{E}_-^S = -E_0 \gamma \frac{k}{\cos\theta_i} \frac{\sin k_z^i d}{k_z^i} [-(\cos\theta_i \widehat{\mathbf{a}}_y + \sin\theta_i \widehat{\mathbf{a}}_z)(\cos\theta_i \widehat{\mathbf{a}}_y - \sin\theta_i \widehat{\mathbf{a}}_z)$$
$$\times S_2(\pi - 2\theta_i) + \widehat{\mathbf{a}}_x \widehat{\mathbf{a}}_x S_1(\pi - 2\theta_i)] \cdot \widehat{\mathbf{e}}_i , \qquad (8)$$

where $S(0) \equiv S_1(\theta = 0) = S_2(\theta = 0)$ is called the forward scattering amplitude, $\gamma \equiv 3f/2x^3$, $x \equiv ka$ is the size parameter, $f = N4\pi a^3/3V$ is the filling fraction of spheres, $\pi - 2\theta_i$ is the specular direction, and we recall that $k_z^i = k\cos\theta_i$. Notice that while $\mathbf{E}_+^S$ is directly proportional to d, $\mathbf{E}_-^S$ is proportional to $\sin k_z^i d / k_z^i$. Here d is the thickness of the averaging region where the centers of the spheres are randomly located. Since we are considering that the slab is thin enough for the independent scattering approximation to be valid, one can take d small enough and approximate $\sin k_z^i d / k_z^i \approx d$.

Notice also, that in general $|\mathbf{E}_+^S| \neq |\mathbf{E}_-^S|$, and this is a direct consequence of the forward–backward anisotropy of Mie scattering, that is, $S(0) \neq S_m(\pi - 2\theta_i)$ for $m = 1, 2$. We also recall that this anisotropy is more acute the larger the sphere. For spheres whose radii are very small with respect to the incident wavelength, this anisotropy almost disappears and one has $|\mathbf{E}_+^S| \approx |\mathbf{E}_-^S|$.

Now the idea is to find the effective current distribution that act as a source of these fields, and identify this effective currents with the average current distribution induced in an effective medium. To model this effective currents within the thin slab, we imagine the simplest possible geometry: a 2D homogeneous and isotropic sheet with no internal structure. We locate the sheet at the $z = 0$ plane and consider an incident plane wave with TE polarization: $\mathbf{E}^i(\mathbf{r}, t) = E_0 \exp\left[i(k_y^i y + k_z^i z)\right] \hat{\mathbf{a}}_x$. The fields radiated by this 2D-sheet of homogeneous material can be found by assuming some 2D–currents (i.e., surface currents) driven by the incident field and applying Maxwell's equation in the region about $z = 0$.

The radiated fields by the 2D–sheet are found in the form

$$\mathbf{E}^J = \begin{cases} E_+^J \exp(i\mathbf{k}^i \cdot \mathbf{r})\hat{\mathbf{a}}_x & \text{for } z > 0 \\ E_-^J \exp(i\mathbf{k}^r \cdot \mathbf{r})\hat{\mathbf{a}}_x & \text{for } z < 0, \end{cases} \tag{9}$$

where $\mathbf{k}^i$ and $\mathbf{k}^r$ have the same meaning as before. The coefficients E_+^J and E_-^J are found in terms of the effective currents. Then these currents are assumed proportional to the incident field through some effective optical coefficients.

If we assume only open currents along the direction of the electric field, $\mathbf{J} = j_{0x}\delta(z)\exp(ik_y^i y)\hat{\mathbf{a}}_x$, we get $E_{\pm x}^J = -\frac{1}{2}\mu_0 j_{0x}\omega/k_z^i$ where μ_0 is the magnetic permeability of vacuum. The resulting radiated field are similar to the ones radiated by the slab with spherical inclusions. However, in this case $E_{+x}^J = E_{-x}^J$, while in the case of the slab one has a right–left anisotropy, that is $E_{+x}^S \neq E_{-x}^S$, which comes from the anisotropy of Mie scattering. Furthermore, the result $E_{+x}^J = E_{-x}^J$ is a direct consequence of Faraday's law $\nabla \times \mathbf{E} = -i\omega\mathbf{B}$, that demands the continuity of E_x^J whenever B_y^J is finite. Here $\mathbf{B}$ is the magnetic field of the incident plane wave. Therefore, if one wants to find a distribution of induced currents that properly simulate the sources of the fields radiated by the slab, one is forced to conclude that this is not possible with an open current distribution. The fulfillment of Faraday's law requires a singular value of B_y at $z = 0$, as the only way to obtain a right-left anisotropy in the wave amplitudes of the radiated electric field. But the only way to get a singular value of B_y at $z = 0$ would be to have a distribution of closed currents that generate a magnetization $\mathbf{M}$ in the sheet along the y–direction. Only in this manner $B_y/\mu_0 = H_y + M_y$ can have a singular contribution. An average of closed currents running along the x–direction can be written as two surface current densities running in opposite directions These closed currents should be induced by an electric field generated by the time variations of the magnetic field along the y–direction. In a slab with spherical inclusions the closed currents can be induced at the inclusions. Let us now define the magnetization field $\mathbf{M}$ as $\mathbf{J} = \nabla \times \mathbf{M}$ where $\mathbf{J}$ is, in general, the average of the closed currents induced in the material. The magnetization in the y–direction can be written as $\mathbf{M} = m_{0y}\delta(z)\exp(ik_y^i y)\hat{\mathbf{e}}_y$, where m_{0y} is the surface magnetization. Now, one can show that the electric field radiated by this induced magnetization is also given by plane waves, as the ones in Eq. (9), but with amplitudes $E_{\pm x}^J = \pm\frac{i}{2}\omega\mu_0\, m_{0y}$, which are discontinuous at $z = 0$. This discontinuity obviously arises from the discontinuity of the closed–current distribution. However, if H_y can induce closed currents in the sheet, the same should happen with the time variations of H_z. In this case closed currents should be induced in the XY plane with a corresponding magnetization in the z–direction. Therefore in order to be consistent we should also consider the field radiated by a source like $\mathbf{M} = -m_{0z}\delta(z)\exp(ik_y^i y)\hat{\mathbf{e}}_z$. Adding up the contributions to the amplitude of the radiated field of the three sources, one gets

$$E_{\pm x}^J = \frac{1}{2}\mu_0\omega\left[-\frac{j_{0x}}{k_z^i} \pm im_{0y} + im_{0z}\frac{k_y^i}{k_z^i}\right]. \tag{10}$$

We now assume that the averages of the induced currents are proportional to the incident field through some effective response functions, and then try to find out the values for which one recovers the fields radiated by the thin slab with spherical inclusions. First we define the polarization field $\mathbf{P}$ as $\mathbf{J} = \partial\mathbf{P}/\partial t \rightarrow -i\omega\mathbf{P}$, where $\mathbf{J}$ is the average of the induced current in the material. Then we define the electric susceptibility tensor $\overline{\overline{\chi}}^E$ as $\mathbf{P} = \epsilon_0\overline{\overline{\chi}}^E \cdot \mathbf{E}$, where is $\mathbf{E}$ the average electric field. In the same manner the magnetic susceptibility tensor $\overline{\overline{\chi}}^H$ is defined as $\mathbf{M} = \overline{\overline{\chi}}^H \cdot \mathbf{H}$, where $\mathbf{H}$ is the average H–field. For an object like the 2D sheet we can write $\overline{\overline{\chi}}_S^E = (\chi_{S\parallel}^E, \chi_{S\parallel}^E, \chi_{S\perp}^E)$ and $\overline{\overline{\chi}}_S^H = (\chi_{S\parallel}^H, \chi_{S\parallel}^H, \chi_{S\perp}^H)$, where the subindexes $\parallel$ and $\perp$ denote parallel and perpendicular to the sheet. The response of the 2D sheet is clearly anisotropic in the $\parallel$ and $\perp$ directions, but we are regarding the XY plane isotropic. Now we assume that the system is so dilute that the average induced current and magnetization distributions are proportional to the *incident* field, thus may write

$$j_{0x} = -i\omega\epsilon_0\chi_{S\parallel}^E E_0 \tag{11}$$

$$m_{0y} = \chi_{S\parallel}^H H_0\cos\theta_i = \chi_{S\parallel}^H\frac{k}{\omega\mu_0}E_0\cos\theta_i \tag{12}$$

$$m_{0z} = \chi_{S\perp}^B B_0\sin\theta_i = \chi_{S\perp}^B\frac{k}{\omega}E_0\sin\theta_i\,, \tag{13}$$

where we have used the relations between $\mathbf{E}$, $\mathbf{H}$ and $\mathbf{B}$ given by Maxwell's equations and we have introduced in Eq. (13) the surface response $\chi_{S\perp}^B$

to the **B** field instead of the response $\chi_{S\perp}^{H}$ to the **H** field. We do this because in case the magnetization is along the z–direction, the field $H_z = B_z/\mu_0 - m_{0z}\delta(z)\exp(ik_y^i y)$ is singular at $z = 0$, and it is not adequate to define a response to a singular field. On the contrary, the field B_z is continuous and can be regarded as the driving field of the induced magnetization.

Now we compare the amplitudes of the waves radiated by this sheet, characterized by three effective surface response functions, with the amplitudes of the waves radiated by the slab with spherical inclusions. In order to do this we first imagine that the effective response of the sheet is actually describing the response of a slab of a finite width d. One can regard the sheet as the ending shape of a limiting process which starts with a slab of a finite width. For example, one can define the surface susceptibility $\chi_{S\parallel}^{E}$ as $\chi_{S\parallel}^{E} = \lim_{d\to 0}\chi^{E}d$, where χ^{E} is the bulk susceptibility of a homogeneous and isotropic slab. Therefore, we have to perform, in Eqs. (11)-(13), the following replacements: $\chi_{S\parallel}^{E} \to \chi^{E}d$ and $\chi_{S\parallel}^{H} \to \chi^{H}d$, where χ^{H} is the bulk magnetic susceptibility of a homogeneous and isotropic slab, and $\chi_{S\perp}^{B} \to \chi^{H}d/\mu \approx \chi^{H}d/\mu_0$. In this last replacement we are taking into account that in the $\perp$ direction there is a surface magnetization at the two parallel faces of the slab that produces a difference between the average **B** and **H** field. This does not happen along the $\parallel$ direction because along this direction the system is boundless. Nevertheless, since we are considering here only the dilute limit, in which the driving field for the induced currents comes solely from the incident beam, we can take $\mathbf{B} \approx \mu_0\mathbf{H}$ and replace $\chi_{S\perp}^{B} \to \chi^{H}d/\mu_0$. We now substitute the replacements in Eqs. (11)–(13) into Eq. (10) and compare it with Eqs. (7) and (8) to yield

$$\chi^{E} + \chi^{H}\cos^2\theta_i + \chi^{H}\sin^2\theta_i = 2i\gamma S(0) \tag{14}$$

$$\chi^{E} - \chi^{H}\cos^2\theta_i + \chi^{H}\sin^2\theta_i = 2i\gamma S_1(\pi - 2\theta_i)\frac{\sin k_z^i d}{k_z^i d}, \tag{15}$$

If we assume $k_z^i d \ll 1$, we can approximate $\sin k_z^i d/k_z^i d \approx 1$. We now solve Eqs. (14) and (15) for χ^{E} and χ^{H} and use the definitions of the electrical permittivity $\tilde{\epsilon} \equiv \epsilon/\epsilon_0 = 1 + \chi^{E}$ and the magnetic permeability $\tilde{\mu} \equiv \mu/\mu_0 = 1 + \chi^{H}$ to get

$$\tilde{\mu}_{eff}^{TE}(\theta_i) = 1 + i\gamma\frac{S_{-}^{(1)}(\theta_i)}{\cos^2\theta_i} \tag{16}$$

$$\tilde{\epsilon}_{eff}^{TE}(\theta_i) = 1 + i\gamma\left[2S_{+}^{(1)}(\theta_i) - S_{-}^{(1)}(\theta_i)\tan^2(\theta_i)\right], \tag{17}$$

where $S_+^{(m)}(\theta_i) \equiv \frac{1}{2}\left[S(0) + S_m(\pi - 2\theta_i)\right]$ and $S_-^{(m)}(\theta_i) \equiv S(0) - S_m(\pi - 2\theta_i)$, and we have added, to $\tilde{\epsilon}$ and $\tilde{\mu}$, the superindex TE to denote the polarization and the subindex eff to emphasize the fact that they describe an effective response. In the case of TM polarization one performs an analogous procedure as the one developed above for TE polarization, and one can show that the corresponding optical coefficients are given by

$$\tilde{\epsilon}_{eff}^{TM}(\theta_i) = 1 + i\gamma \frac{S_-^{(2)}(\theta_i)}{\cos^2\theta_i} \tag{18}$$

$$\tilde{\mu}_{eff}^{TM}(\theta_i) = 1 + i\gamma \left[2S_+^{(2)}(\theta_i) - S_-^{(2)}(\theta_i)\tan^2(\theta_i)\right]. \tag{19}$$

These results can also be readily obtained from the symmetry relations in Maxwell's equations.

Note that the effective optical coefficients $\tilde{\epsilon}_{eff}$ and $\tilde{\mu}_{eff}$ depend on the angle of incidence and on the polarization, just like in an anisotropic medium. Also, the expressions for the effective optical coefficients in Eqs. (16)-(19) are linear in $\gamma \equiv 3f/2x^3$ and they are valid only to linear order in γ. This is consistent with the dilute–limit approximation adopted above, and therefore the validity of all of our results will be limited by this restriction.

According to continuum electrodynamics the effective index of refraction n_{eff} should be given by

$$n_{eff}^{(m)}(\theta_i) = \sqrt{\tilde{\epsilon}_{eff}^{(m)}(\theta_i)\tilde{\mu}_{eff}^{(m)}(\theta_i)} \simeq \sqrt{1 + 2i\gamma S(0)}, \tag{20}$$

where we dropped terms of second order in γ since our approximations are valid up to first order in γ only. This result is the same as the one derived by Foldy [16] long time ago We may expand the square root and to lowest order in γ we get $n_{eff} \approx 1 + i\gamma S(0)$ which is isotropic and independent of polarization, and is actually the same result as the one proposed by van de Hulst [3]. So we can see that although the optical coefficients $\tilde{\epsilon}_{eff}$ and $\tilde{\mu}_{eff}$ are highly anisotropic and polarization dependent, their dependence in the angle of incidence is such that the square root of their product is not.

Let us now look at some limiting cases. First we notice that for small particles ($x \ll 1$) the Mie forward–backward anisotropy in the angular distribution of scattered radiation becomes $S_1(\theta_i) \approx -ix^3\beta$ and $S_2(\theta_i) \approx -ix^3\beta\cos\theta_i$, where $\beta = (\tilde{\epsilon}_S - 1)/(\tilde{\epsilon}_S + 2)$ and $\tilde{\epsilon}_S = \epsilon_S/\epsilon_0$ is the electrical permittivity of the spheres. Then $S_+^{(1)} \approx -ix^3\beta$, $S_-^{(1)} \approx 0$, $S_+^{(2)} \approx -ix^3\beta\sin^2\theta_i$, and $S_-^{(2)} \approx -2ix^3\beta\cos^2\theta_i$. Substituting these values into Eqs. (16)-(19) we get

$$\tilde{\mu}_{eff}^{TE}(\theta_i) = \tilde{\mu}_{eff}^{TM}(\theta_i) \equiv \tilde{\mu}_{eff} = 1 \tag{21}$$

$$\tilde{\epsilon}_{eff}^{TE}(\theta_i) = \tilde{\epsilon}_{eff}^{TM}(\theta_i) \equiv \tilde{\epsilon}_{eff} = 1 + 3\beta f, \tag{22}$$

and these are the well-known results for the case of small particles, or for the case of an ordinary material, when one regards the material as a composite made of molecular inclusions in vacuum. Eq. (21) tells us that the system is non magnetic and Eq. (22) is the low–density limit of the effective dielectric response in Maxwell–Garnett theory or in the Clausius–Mossotti relation, when one interprets β as proportional to the molecular polarizability. One can also see that the magnetic character of the system appears only when the spheres are big enough and is related to the large forward–backward anisotropy in the Mie scattering of large particles ($x \sim 1$).

For normal incidence ($\theta_i = 0$) one gets

$$\tilde{\mu}_{eff}^{TE}(0) = \tilde{\mu}_{eff}^{TM}(0) \equiv \tilde{\mu}_{eff}(0) = 1 + i\gamma\left[S(0) - S_1(\pi)\right] \tag{23}$$

$$\tilde{\epsilon}_{eff}^{TE}(0) = \tilde{\epsilon}_{eff}^{TM}(0) \equiv \tilde{\epsilon}_{eff}(0) = 1 + i\gamma\left[S(0) + S_1(\pi)\right], \tag{24}$$

where we have used $S_1(\pi) = -S_2(\pi)$. And these are the results proposed by C. Bohren [6] when he introduced the idea of a magnetic response in the optical properties of granular materials made with non–magnetic components.

At grazing incidence, $\theta_i \to \pi/2$, we have that $S_m(\pi-2\theta_i) \to S(0)$, thus $S_+^{(m)}(\theta_i) \to S(0)$ and $S_-^{(m)}(\theta_i) \to 0$ and $S_-^{(m)}(\theta_i)/\cos^2\theta_i$ remains finite. We can see this by expanding $S_-^{(m)}(\theta_i)$ around $\theta_i = \pi/2$ and showing that $\lim_{\theta_i \to \pi/2} S_-^{(m)}(\theta_i)/\cos^2\theta_i = 2S_m''(0)$, where the primes indicate derivative with respect to the argument.

If we now accept the description of the optical properties of a granular material in terms of the effective optical coefficients given by Eqs. (16)-(19), the reflection amplitudes of a half space r_{hs} will be given by the Fresnel's relations of continuum electrodynamics, that is,

$$r_{hs}^{TE} = \frac{\tilde{\mu}_{eff}^{TE}(\theta_i)k_z^i - k_z^{eff}}{\tilde{\mu}_{eff}^{TE}(\theta_i)k_z^i + k_z^{eff}} \quad \text{and} \quad r_{hs}^{TM} = \frac{\tilde{\epsilon}_{eff}^{TM}(\theta_i)k_z^i - k_z^{eff}}{\tilde{\epsilon}_{eff}^{TM}(\theta_i)k_z^i + k_z^{eff}}, \tag{25}$$

where $k_z^{eff} = k\sqrt{(n^{eff})^2 - \sin^2\theta_i}$, and $n^{eff} = 1 + i\gamma S(0)$. One can see that these reflection amplitudes look very different from the ones we would have used by assuming a non–magnetic effective medium with $\tilde{\epsilon}_{eff} = n_{eff}^2 = [1 + i\gamma S(0)]^2$ and $\tilde{\mu}_{eff} = 1$. We will denote the reflection coefficients calculated by r_{n-m}^{TE} and r_{n-m}^{TM} where the subscript n-m stands for *non–magnetic*.

Now, it is also possible to derive the coherent reflectance from the scattering approach by considering a semi–infinite pile of slabs with spherical particles and solving the multiple scattering of waves between slabs. The result from this approach can be shown to be consistent with Eq. (25) [14]. Extending the above results to a composite matrix consisting of spherical inclusions embedded in a homogeneous matrix is not difficult and is also discussed in Ref. [14].

4. NUMERICAL RESULTS

Now, we illustrate the behavior of the optical coefficients in a few examples by numerical calculations. We plot the normalized change in the optical coefficients (real and imaginary parts). By *normalized* we mean divided by the fractional volume occupied by the spheres f and by *change* in the optical coefficient we mean the difference with respect to the optical coefficient without the particles (vacuum, in this case). In all cases we assume that the particles are in vacuum (no matrix) and we should recall that the expressions used are valid for dilute systems only ($f \ll 1$). The scattering matrix elements S, involved in the formulas for the optical coefficients were calculated following the recipe given in the book by Bohren and Hoffman [15].

First, in Fig. 2, we show the effective index of refraction for a system of non–magnetic lossless glass spheres ($n_p = 1.50$) as a function of the particle radius divided by the wavelength, and similar plots for lossy spheres with increasing imaginary component of the refractive index. As it can be appreciated, even if the spheres are lossless the effective index of refraction has an imaginary component. This is entirely due to scattering losses and has a maximum near $a/\lambda \simeq 0.5$. For curves corresponding to lossy particles ($\Im(n_p) \neq 0$) the loss is due to both, absorption in the particles, and scattering from the particles. Also, note that the real part of the effective index of refraction can be less than one for some particle radius. As it may be expected, the imaginary part of the effective index of refraction reach the highest value for most absorbing particles. However, also the real part of the effective refractive index is higher than for the other curves. The reason is that the scattering efficiency is stronger for these particles, since the contrast of the particles with respect to vacuum is highest.

In Fig. 3 we plot the real and imaginary part of the normalized change in the effective optical coefficients, $\epsilon_{eff}^{TE}, \epsilon_{eff}^{TM}, \mu_{eff}^{TE}$, and μ_{eff}^{TM} as a function of the particle radius divided by the wavelength for a system of

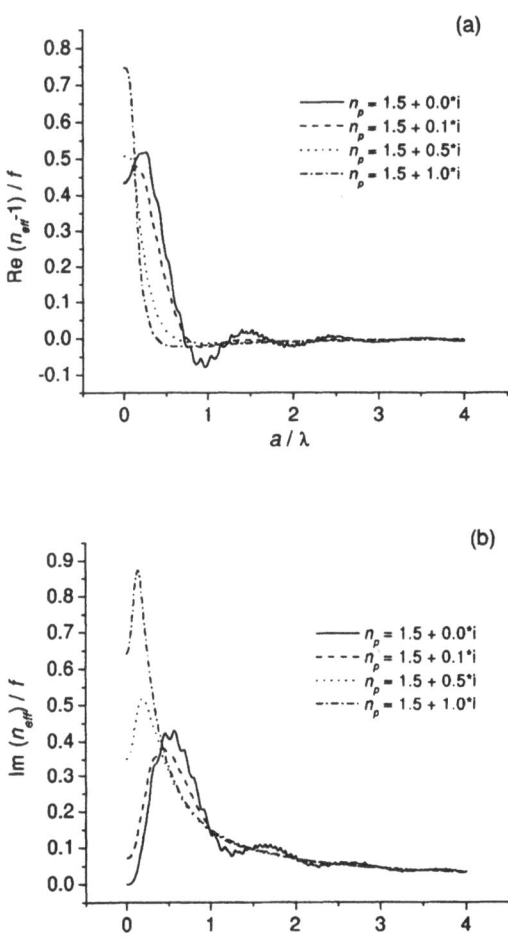

Figure 2. Plots of the normalized change in the real and imaginary part of the effective index of refraction for a system of non–magnetic glass spheres ($n_p = 1.50$) in vacuum, and similar plots for particles with different values of the imaginary part of their index of refraction.

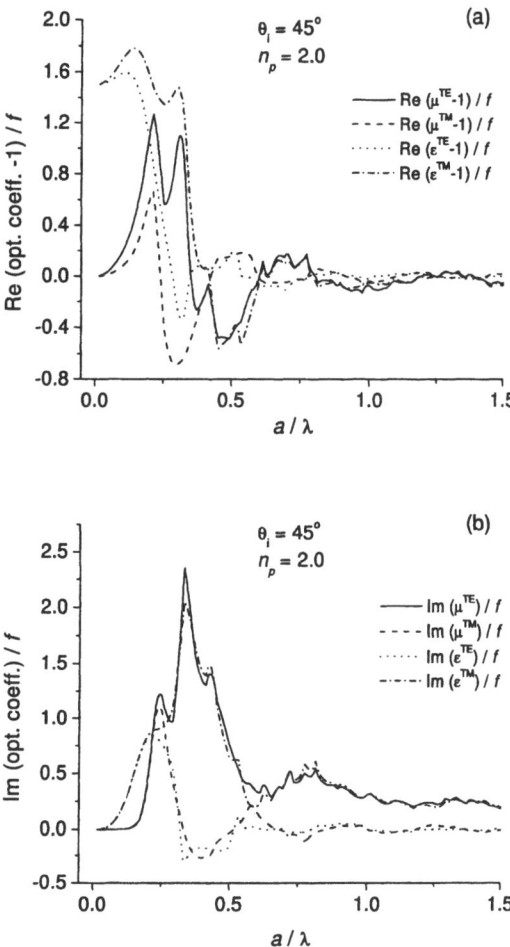

Figure 3. Plots of the normalized change in the real and imaginary part of the optical coefficients as a function of the particle radius a divided by the wavelength λ, for an angle of incidence of 45°. The plots are for a system of non–magnetic dielectric spheres ($n_p = 2.00$). The subindex in the optical coefficients *eff* was removed here for clarity; dot–dot lines are for ϵ^{TE}, dash–dot lines for ϵ^{TM}, solid lines for μ^{TE}, and dash–dash lines for μ^{TM}.

dielectric spheres with refractive index of $n_p = 2.0$. and for an angle of incidence of $\theta_i = 45°$. As it can be appreciated, these are irregular oscillatory function of the particle radius. Also, the effective magnetic permeability reaches values comparable to the effective electric permittivity for particles of radius $a/\lambda \sim 0.25$ and larger. In Fig. 3b it can be seen that the imaginary parts of ϵ_{eff}^{TE} and μ_{eff}^{TM} are negative within some range of particle radius. This, however, is not an inconsistency since the sums $\Im\epsilon_{eff}^{TE} + \Im\mu_{eff}^{TE}$, and $\Im\epsilon_{eff}^{TM} + \Im\mu_{eff}^{TM}$, remain always positive. In similar plots, but for fixed particle radius and as a function of the angle of incidence (not shown here) one finds that the change in the optical coefficients generally increases towards grazing incidence.

Since the appearance of the effective magnetic susceptibility for systems of non–magnetic particles is apparently due to the induced closed currents within the particles, it is interesting to compare the magnetic response of systems of dielectric particles with that of metallic particles. It turns out that the effective magnetic response for systems of particles of the same radius is in general similar in magnitude for dielectric and metallic particles; except when the particle radius is small compared to the wavelength. For small particles (say, $a \lesssim 0.1\lambda$) the imaginary component of the effective magnetic permeability is orders of magnitude larger for metallic particles than for dielectric ones. Although its value is small in absolute terms. Also, the change in the real part of μ_{eff} is negative for metallic particles and positive for dielectric particles. In Fig. 4 we plot the real and imaginary parts of the normalized change in μ_{eff} for both polarizations as a function of the angle of incidence for metallic particles (copper, $n_p = 0.21 + 4.05i$ at $\lambda = 0.69\,\mu m$) and dielectric (glass, $n_p = 1.5$) ones, and for particles radius of $a/\lambda = 0.1$.

Finally, with respect to the coherent reflectance, we have found that this is generally smaller than what would be predicted by a simpler model which ignores the effective magnetic response. In Fig. 5 we show the reflectance for TE and TM polarizations as a function of the angle of incidence for particles with $n_p = 2.50$ and two different radius: $a/\lambda = 0.1$ and $a/\lambda = 0.5$. The fractional volume of the particles is taken to be $f = 0.1$. In Fig. 5b the location of the Brewster angle can be appreciated and it can be seen that the location of the Brewster angle predicted by the non–magnetic model differs from our result. The curves for TM polarization and for particles $a/\lambda = 0.1$ are an exception and the reflectance predicted by the non–magnetic formula is lower than the coherent reflectance R. In plots of the TE reflectance for larger particles (not shown here) one finds zeros in the coherent reflectance, and these can be interpreted as a Brewster angle, which only exist when the medium has a magnetic permeability different from that of vacuum.

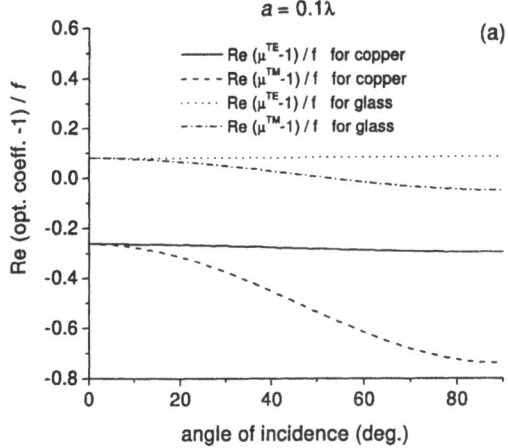

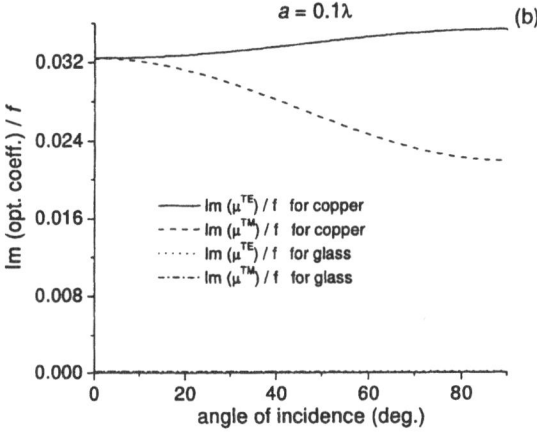

Figure 4. Plots of the normalized change in the (a) real and (b) imaginary part of the effective magnetic permeability as a function of the angle of incidence for metallic (copper) and glass particles of radius $a = 0.1\lambda$. Plots for both polarizations are shown. The subindex in the optical coefficients *eff* was removed here for clarity; dot–dot lines are for μ^{TE} and dash–dot lines for μ^{TM}, both for copper particles. Solid lines for μ^{TE} and dash–dash lines are for μ^{TM}, both for glass particles.

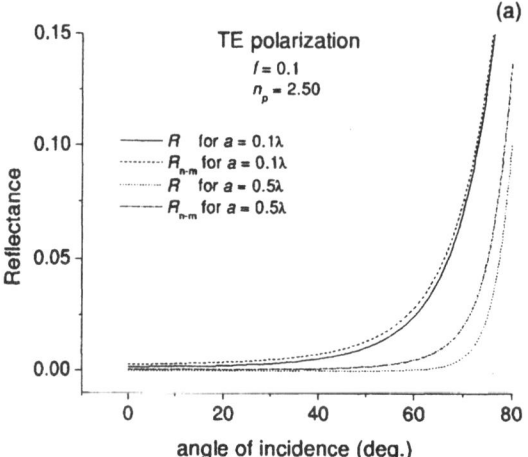

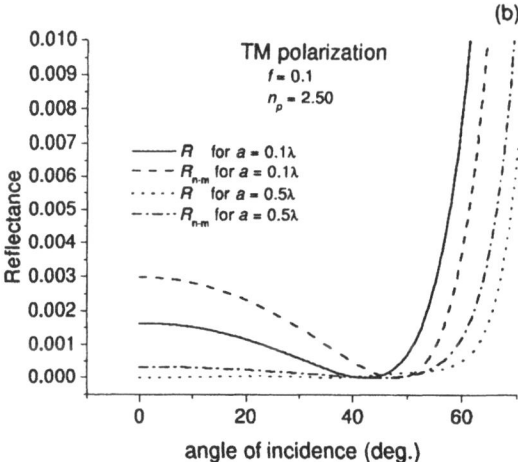

Figure 5.   Plot of the coherent reflectance R of polarized light for a system of non-magnetic dielectric spheres of index of refraction $n_p = 2.50$ and a filling fraction of $f = 0.1$; (a) TE polarization and (B) TM polarization. For comparison we also plot the reflectance ignoring the effective magnetic susceptibility ($R_{n\text{-}m}$).

5. CONCLUSIONS

We have constructed an effective medium theory to describe the coherent reflection of electromagnetic waves from a random system of spheres. Our results can be regarded as an extension of ideas put forth previously by C. Bohren. We found that the effective medium must posses an effective magnetic permeability, even if the particles are non–magnetic, in order to have a theory consistent with continuum electrodynamics. The effective magnetic susceptibility becomes comparable to the effective electric permittivity as the particle radius increases and they attain their maximum value when the radius is comparable to the wavelength of the incident radiation. The origin of this magnetic effect appears in our theory from the identification of induced closed currents as sources of the fields radiated by the random system of spheres. These closed currents must be physically present within each sphere and when averaged they must act as the source of an effective magnetization. The coherent reflectance calculated including the effective magnetic response differs appreciably from the one calculated without it. The formulas put forth in this work are valid for a dilute system of spheres and they can be readily used in applications satisfying this criteria. Extensions of the present results to a polydispersed system of spheres is straightforward.

Acknowledgments

We acknowledge the partial support of Consejo Nacional de Ciencia y Tecnología (México) and Centro de Investigación en Polímeros (Grupo COMEX, Mexico) through Grant RI–1200–2–1. The partial support of Dirección General del Personal Académico of Universidad Nacional Autónoma de México (Mexico) through Grant IN–104201 is also acknowledged.

References

[1] See, for example, the historical review of R. Landauer, in: *Proceedings of the First Conference on Electrical Transport and Optical Properties of Inhomogeneous Media*, AIP Conf. Proc. No. 40, J.C. Garland and D. B. Tanner, eds. (AIP, New York, 1978) pp. 2. Also, for more recent references, see, for example, *Proceedings of the Fifth International Conference on Electrical Transport and Optical Properties of Inhomogeneous Media (ETOPIM 5)*, P.M. Hui, P. Sheng and L.-H. Tang, eds. Physica **B279** (2000) Nos. 1–3.

[2] See, for example, P. Chýlek, G. Videen, D.J.W. Geldart, J.S. Dobbie and H.C.W. Tso, in: *Light Scattering by Nonspherical Particles*, M.I. Mishchenko, J.W. Hovenier and L. D. Travis, eds. (Academic Press, 2000).

[3] H.C. van de Hulst, *Light scattering by small particles* (John Wiley & Sons Inc., New York NY, 1957).

[4] L. Tsang and J.A. Kong, *Scattering of Electromagnetic Waves; Advanced Topics* (John Wiley and Sons, Inc., New York N.Y., 2001) pp. 128.

[5] G. Videen and P. Chýlek, Optics Commun. **158** (1998) 1.

[6] C.F. Bohren, J. Atmosph. Sci. **43** (1986) 468.

[7] J.B. Pendry, A.J. Holden, D.J. Robbins and W.J. Stewart, IEEE Trans. Microwave Theory Tech. **47** (1999) 2075.

[8] Y. Kuga, D. Rice, and R.D. West, IEEE Trans. Antenn. Propagat. 44 (1996) 326.

[9] C. Yang, A. Wax, and M.S. Feld, Opt. Lett. **26** (2001) 235.

[10] G.H. Meeten and A.N. North, Meas. Sci. Technol. **6** (1995) 214.

[11] M. Mohammadi, Advances in Colloid and Interface Science **62** (1995) 17.

[12] A. García–Valenzuela, M. Peña–Gomar, and C. Fajardo–Lira, Opt. Eng. (2002) in press.

[13] W.L. Mochán and R.G. Barrera, Phys. Rev. **B32** (1985) 4984.

[14] R.G. Barrera and A. García–Valenzuela, "Coherent reflectance in a system of random Mie scatterers: An effective-medium approach", submitted for publication.

[15] C. F. Bohren and D.R. Huffman, *Absorption and Scattering of Light by Small Particles* (John Wiley & Sons, New York N.Y., 1983).

[16] L.L. Foldy, Phys. Rev. **67** (1945) 107.

SPIN AND CHARGE ORDER IN THE VORTEX LATTICE OF THE CUPRATES: EXPERIMENT AND THEORY

Subir Sachdev

Department of Physics, Yale University,
P.O. Box 208120, New Haven CT 06520-8120, USA
Email:subir.sachdev@yale.edu; Web:http://pantheon.yale.edu/~subir

Abstract   I summarize recent results, obtained with E. Demler, K. Park, A. Polkovnikov, M. Vojta, and Y. Zhang, on spin and charge correlations near
a magnetic quantum phase transition in the cuprates. Static charge order coexisting with dynamic spin correlations has recently been observed
around vortices in slightly overdoped $Bi_2Sr_2CaCu_2O_{8+\delta}$ (J. E. Hoffman
et al., Science **295**, 466 (2002)), and neutron scattering experiments
have measured the magnetic field dependence of static spin order in the
underdoped regime in $La_{2-\delta}Sr_\delta CuO_4$ (B. Lake *et al.*, Nature **415**, 299
(2002)) and $LaCuO_{4+y}$ (B. Khaykovich *et al.* cond–mat/0112505). Our
predictions provide a semi–quantitative description of these observations, with only a single parameter measuring distance from the quantum critical point changing with doping level. These results suggest
that a common theory of competing spin, charge and superconducting
orders provides a unified description of all the cuprates.

Keywords: Spin density wave, charge density wave, superconductivity, vortex lattice

1.   INTRODUCTION

Three recent experiments [1, 2, 3] have shed new light on the spin and
charge density wave collective modes of the cuprate superconductors.
This article will summarize the main results of our theory [4, 5, 6, 7,
8, 9, 10] of these collective modes in the vicinity of a quantum phase
transition between two superconducting states, only one of which has
static, long–range, spin density wave order; we will also connect our
theory to these experiments. In particular, we suggested [6] that static
charge order should coexist with dynamic spin–gap fluctuations around
the vortex cores in the cuprate superconductors, as has been observed

Developments in Mathematical and Experimental Physics,
Volume B: Statistical Physics and Beyond
Edited by Macias *et al.*, Kluwer Academic/Plenum Publishers, 2003

in [1]. We also discussed [5] a singular field dependence for the static magnetic moment in the underdoped cuprates, and this is consistent with [2, 3].

The starting hypothesis of our theory is that the collective spin excitations of the doped cuprates can be described by using the proximity of a magnetic quantum critical point. This was proposed in Ref. [11]; almost simultaneously, NMR experiments in $La_{2-\delta}Sr_\delta CuO_4$ [12] showed crossovers which could be neatly interpreted in terms of a magnetic quantum critical point near a doping concentration $\delta \approx 0.12$ with dynamic exponent $z = 1$. The ground state is a good superconductor at this value of δ, and so the magnetic transition takes place between two superconducting states. Evidence supporting this interpretation also appeared in neutron scattering measurements [13]. An explicit theory for a quantum transition between a d–wave superconductor with co–existing long–range spin density wave order (a SC+SDW state) and an ordinary d–wave superconductor (a SC state) was first discussed by Balents *et al.* [14]; they focused on the case where the SDW ordering wavevector was exactly equal to the spacing between the two nodal points where the d–wave superconductor has gapless quasi–particle excitations, and studied their role in the critical theory. However, their analysis also makes it clear that the nodal quasiparticles can be safely neglected for the generic case in which the wavevector matching condition is not satisfied [4], and we will mainly discuss this simpler case here. The SC+SDW to SC transition in this case is formally identical to that in an insulator, and the SC state has a sharp $S = 1$ 'resonance peak' associated with stable $S = 1$ collective excitonic excitation [15]. Such a theory for the SC to SC+SDW transition was used by us [16] to predict the effects of Zn impurities on the resonance peak in the SC phase.

2.   ORDER PARAMETER AND FIELD THEORY

Neutron scattering experiments show that the lowest energy collective spin excitations at and above $\delta \approx 0.12$ reside near the wavevectors $\mathbf{K}_x = (3\pi/4, \pi)$ and $\mathbf{K}_y = (\pi, 3\pi/4)$ (our unit of length is the square lattice spacing). So we write for the spin operator at the site $\mathbf{r}$:

$$S_\alpha(\mathbf{r}, \tau) = \mathrm{Re}\left[e^{i\mathbf{K}_x \cdot \mathbf{r}} \Phi_{x\alpha}(\mathbf{r}, \tau) + e^{i\mathbf{K}_y \cdot \mathbf{r}} \Phi_{y\alpha}(\mathbf{r}, \tau) \right], \qquad (1)$$

where $\alpha = x, y, z$ extends over the directions in spin space, and $\Phi_{x,y\alpha}$ are *complex* fields which will serve as order parameters for the SC+SDW to SC transition. In the SC+SDW phase, $\Phi_{x,y\alpha}$ are condensed and the condensate describes the spatial modulation of the spin, by (1); in the SC phase, the $\Phi_{x,y\alpha}$ are dynamically fluctuating, and its quanta are *spin*

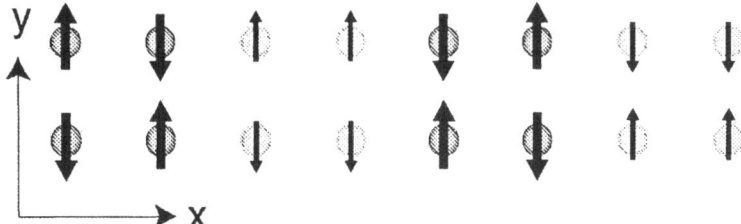

Figure 1. A bond–centered, collinear SDW at the wavevector $(3\pi/4, \pi)$. The size of the arrow represents the mean spin moment, while the shading of the circle is the electron density. The sliding degree of freedom corresponds to a shift in the position of the light and dark circles.

excitons. The representation (1) can describe a large variety of spin modulations *e.g.* the state with $\langle \Phi_{x\alpha} \rangle \propto (1, i, 0)$, $\langle \Phi_{y\alpha} \rangle = 0$ is a spiral SDW in the x direction. Experimentally, however, it is clear that the SDW is not spiral, but *collinear*; an example of a collinear SDW has $\langle \Phi_{x\alpha} \rangle \propto e^{i\theta}(1, 0, 0)$, $\langle \Phi_{y\alpha} \rangle = 0$—notice that the average spin vectors on all sites are parallel or antiparallel. The phase θ represents a *sliding* degree of freedom of the SDW: for the special commensurate value of $\mathbf{K}_x$ under consideration here, the coupling to the lattice will prefer that θ take one of the values $n\pi/4$ (a site–centered SDW) or $(n + 1/2)\pi/4$ (a bond–centered SDW) where $n = 0 \dots 7$ integer. A sketch of a bond–centered SDW is shown in Fig 1. Notice that the magnitude of the spin changes from site to site, and this implies [17] that there must be a corresponding modulation in the electron density, $\delta\rho(\mathbf{r})$: the latter has the same quantum numbers as $\sum_\alpha S_\alpha^2$, and hence we deduce that the charge density wave (CDW) order can be written as

$$\delta\rho(\mathbf{r}, \tau) \propto \text{Re} \sum_\alpha \left[e^{i2\mathbf{K}_x \cdot \mathbf{r}} \Phi_{x\alpha}^2(\mathbf{r}, \tau) + e^{i2\mathbf{K}_y \cdot \mathbf{r}} \Phi_{y\alpha}^2(\mathbf{r}, \tau) \right] + \dots. \quad (2)$$

The period of the CDW is half that of the SDW, and the amplitude of the CDW vanishes for the spiral SDW.

We are interested here in the quantum transition from the SC+SDW state with $\langle \Phi_{x,y\alpha} \rangle \neq 0$ to the SC state with $\langle \Phi_{x,y\alpha} \rangle = 0$ in a background of quiescent superconductivity. The allowed terms in the effective action are constrained by the underlying symmetries: these were discussed in some generality in [9]. Here we will be satisfied by considering a simplified effective action which is written most easily in terms of the real and imaginary components of $\Phi_{x,y\alpha}$:

$$\begin{array}{llll}
\Phi_{xx} = \varphi_1 + i\varphi_7 & \Phi_{xy} = \varphi_2 + i\varphi_8 & \Phi_{xz} = \varphi_3 + i\varphi_9 \\
\Phi_{yx} = \varphi_4 + i\varphi_{10} & \Phi_{yy} = \varphi_5 + i\varphi_{11} & \Phi_{yz} = \varphi_6 + i\varphi_{12}
\end{array} \quad (3)$$

The effective action for the real fields φ_μ, $\mu = 1 \ldots 12$ is taken to be

$$S_\varphi = \int d^2r d\tau \left\{ \frac{1}{2} \left[(\partial_\tau \varphi_\mu)^2 + (\nabla_r \varphi_\mu)^2 + s\varphi_\alpha^2 \right] + \frac{u}{2} (\varphi_\mu^2)^2 \right\} \qquad (4)$$

where a summation over the repeated μ index is implied, and we have chosen units so that the velocity of spin waves is unity. Notice that the action S_φ has a large O(12) symmetry of rotations in μ space. This symmetry is present in *all* allowed terms which are quadratic in φ_μ, but is broken by a number of quartic terms [9] which are not displayed in (4). However, all permitted terms in the effective action do respect the sliding symmetry under which $\Phi_{x,y\alpha} \rightarrow e^{in_{x,y}\pi/4}\Phi_{x,y\alpha}$ for integer $n_{x,y}$. The coupling s serves as the tuning parameter which measures distance from the quantum critical point: the SC+SDW phase will appear for $s < s_c$ and the spin–singlet SC phase for $s > s_c$. The dynamic properties of the transition at $s = s_c$ have been investigated in much detail in earlier work [15].

Now we consider the influence of the magnetic field, H, applied perpendicular to the CuO$_2$ layers. This couples most strongly to the 'background' SC order, and so we are forced to consider the response of the superconducting order parameter $\psi(\mathbf{r})$. In suitable units (discussed in [9]), the free energy for $\psi(\mathbf{r})$ can be written in the familiar Ginzburg–Landau form

$$\mathcal{F} = \Upsilon \int d^2r \left[-|\psi|^2 + \frac{1}{2}|\psi|^4 + |(\nabla_\mathbf{r} - i\mathbf{A})\psi|^2 \right]. \qquad (5)$$

where Υ is a parameter measuring the relative contributions of the magnetic and superconducting energies, and $\nabla_\mathbf{r} \times \mathbf{A} = H\hat{\mathbf{z}}$. Notice that we have assumed a τ independent ψ—this is permissible because the SC order is non–critical and its quantum fluctuations can be safely neglected.

Finally, we have to couple the H–response of $\psi(\mathbf{r})$ to the quantum SDW fluctuations. There are two distinct couplings, which have rather different physical consequences. The first, v, is a simple coupling between the magnitudes of the SC and SDW order parameters, chosen with a repulsive sign ($v > 0$) to account for the competition between these orders:

$$S_v = \frac{v}{2} \int d^2r d\tau \varphi_\mu^2(\mathbf{r},\tau)|\psi(\mathbf{r})|^2. \qquad (6)$$

Such a coupling was discussed by Zhang [18]. This coupling will play an important role in determining the shape of the phase boundaries to be described below. The second coupling, unlike v, recognizes the fact that the vortex lattice induced by H breaks translational symmetry, and so the SDW fluctuations should also not be invariant under the 'sliding' symmetry (the term in (6) is invariant under the sliding symmetry). In particular the vortex core radius in the cuprates is only of the order of a few lattice spacings, and the energy of an SDW fluctuation will certainly change depending upon which portion of the CDW (see Fig 1) is centered on a vortex core. In other words, each vortex core, at a position $\mathbf{r}_v$, will prefer a certain phase of the local CDW order parameter $\sum_\alpha \Phi^2_{x,y\alpha}(\mathbf{r}_v, \tau)$. Expanding the CDW order parameter using (3), we deduce the second term which couples the SC and SDW order parameters:

$$
\mathcal{S}_{\text{pin}} = - \sum_{\{\mathbf{r}_v, \psi(\mathbf{r}_v)=0\}} \int d\tau \left\{ \sum_{\mu=1}^{3} \text{Re}\left[\zeta_x \left(\varphi_\mu(\mathbf{r}_v, \tau) + i\varphi_{6+\mu}(\mathbf{r}_v, \tau) \right)^2 \right] \right.
$$
$$
\left. + \sum_{\mu=4}^{6} \text{Re}\left[\zeta_y \left(\varphi_\mu(\mathbf{r}_v, \tau) + i\varphi_{6+\mu}(\mathbf{r}_v, \tau) \right)^2 \right] \right\}. \quad (7)
$$

The complex coupling constants $\zeta_{x,y}$ measure the pinning strength of the phase of the sliding CDW to some preferred value near each vortex core.

We have now defined a well–posed field theoretical problem, which was analyzed in some detail in [9]: describe the dynamic quantum SDW fluctuations associated with the partition function

$$
\mathcal{Z}\left[\psi(\mathbf{r})\right] = \int \mathcal{D}\varphi_\mu(\mathbf{r}, \tau) \exp\left(-\frac{\mathcal{F}}{T} - \mathcal{S}_\varphi - \mathcal{S}_v - \mathcal{S}_{\text{pin}} \right), \quad (8)
$$

where the optimum value of the static SC order $\psi(\mathbf{r})$ is determined by the minimization of $-\ln \mathcal{Z}\left[\psi(\mathbf{r})\right]$ via the solution of the saddle–point equation

$$
\frac{\delta \ln \mathcal{Z}\left[\psi(\mathbf{r})\right]}{\delta \psi(\mathbf{r})} = 0. \quad (9)
$$

Note the highly asymmetric treatment of the SC and SDW orders.

3.   PHASE DIAGRAM

Our primary results for the properties of (8) and (9) are contained in the phase diagram as a function of s and H in Fig 2. The formal solution of these equations also allows solutions in which the SC order vanishes and $\psi(\mathbf{r}) = 0$ everywhere: this leads to the SDW phase and the "Normal" phase. However we do not expect our theory to be accurate such a regime: quantum fluctuations of the SC order parameter are surely important once $\psi(\mathbf{r})$ becomes small, and these have been neglected in our theory. Our results are more precise in the small H region of Fig 2, in the vicinity of the boundary between the SC+SDW and SC

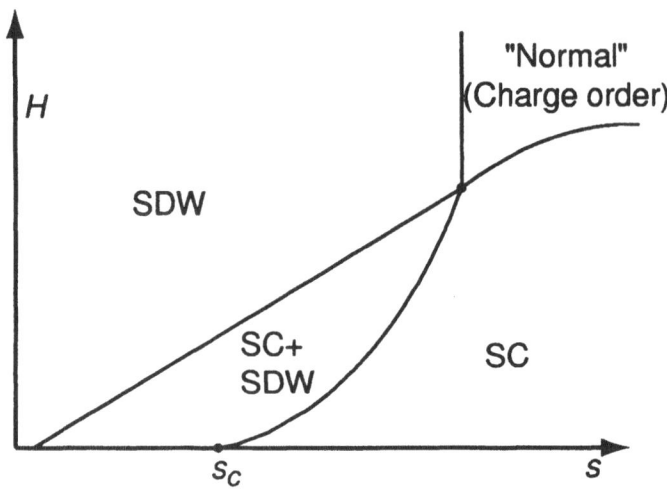

Figure 2.   Zero temperature phase diagram of the model $\mathcal{Z}$ defined in (8) and (9); from Refs. [5] and [9]. The phases have the following expectation values: (*i*) SC: $\langle \varphi_\mu \rangle = 0$, $\langle \psi \rangle \neq 0$, (*ii*) SC+SDW: $\langle \varphi_\mu \rangle \neq 0$, $\langle \psi \rangle \neq 0$, (*iii*) SDW: $\langle \varphi_\mu \rangle \neq 0$, $\langle \psi \rangle = 0$, and (*iv*) "Normal": $\langle \varphi_\mu \rangle = 0$, $\langle \psi \rangle = 0$.

phases; indeed, the functional forms of the main results quoted below are expected to be exact. Upon accounting for the quantum fluctuations of the superconducting order, and the Berry phases associated with the electrons in the nearby Mott insulator [4], we expect that the "Normal" phase will display some sort of static charge order.

An important prediction of our theory is the shape of the phase boundary between the SC and SC+SDW phases at small H. This transition is associated with the condensation of the φ_μ exciton, when the spin gap to the creation of an exciton vanishes in the SC phase. Detailed arguments were presented in [5, 9] showing that this condensation occurs in an exciton state which is extended throughout the entire lattice; indeed, a variational approximation in which the exciton wavefunction is assumed to be simply a constant gives essentially exact results (as opposed to an approximation in which the exciton is strongly localized in the vortex cores [19]). The presence of the vortex lattice in $\psi(\mathbf{r})$ influences the energy of this extended exciton primarily via the v coupling in $\mathcal{S}_v$. When spatially averaged over $\mathbf{r}$, the dominant change in the average value of $|\psi(\mathbf{r})|^2$, and hence in the energy of the exciton, arises from slight suppression of superconductivity in the *superflow* region surrounding each vortex core. The much smaller vortex core region always has a significantly weaker effect on the exciton energy. It is a simple matter to compute the correction to the exciton energy from the superflow, and hence to find the critical field at which the spin gap vanishes. This leads to our result [5, 9] for the phase boundary between the SC and SC+SDW phases:

$$H \sim \frac{(s - s_c)}{\ln(1/(s - s_c))}.$$ (10)

Note that the phase boundary approaches the $H = 0$ limit with vanishing slope: consequently a relatively small field applied to the superconductor for $s > s_c$ will drive the system into the SC+SDW phase. This is our explanation for the shift in the energy of the dynamic spin fluctuations seen in [20].

The phase diagram of Fig 2 leads to a number of predictions [5, 6, 9] for observables in the SC and SC+SDW phases, some of which have been tested in recent experiments [1, 2, 3]. We discuss theory and experiment in the two phases in turn in the following subsections.

3.1. STATIC CHARGE ORDER IN THE SC PHASE

The SC phase has $\langle \varphi_\mu \rangle = 0$, and so the SDW fluctuations are dynamic and the spin exciton only exists above a finite energy gap Δ. The superconducting order $\psi(\mathbf{r})$ is suppressed in the vortex cores, and so this region should exhibit characteristics of the doped spin-gap (*i.e.* paramagnetic) Mott insulator, as was argued in [21]. Paramagnetic Mott insulators, and their response to doping with mobile charge carriers, were studied at some length in [4]: it was argued that the most likely candidates had bond–centered charge order which survived in a superconducting state for a finite range of doping—this work will be reviewed further in Section 4. Reasoning in this manner, Ref. [6] predicted that static charge order should appear in and around the vortex cores, coexisting with dynamic spin fluctuations in the SC state. Order of this type appears to have been seen in the recent STM experiment on $Bi_2Sr_2CaCu_2O_{8+\delta}$ [1].

The spatial extent of the charge order was computed in the field theory in (8), (9) in [9]. The gapped spin exciton, φ_μ, views the vortex lattice as a periodic potential, and consequently its dispersion develops the Bloch structure of a particle moving in a periodic potential. To zeroth order in the pinning terms, $\zeta_{x,y}$, the two-point ϕ_μ Green's function is diagonal in the μ index, and its diagonal component can be written as [5, 9]

$$G_\varphi(\mathbf{r}, \mathbf{r}', \omega_n) = \sum_\mu \int_{1BZ} \frac{d^2 k}{4\pi^2} \frac{\Xi^*_{\mu\mathbf{k}}(\mathbf{r}) \Xi_{\mu\mathbf{k}}(\mathbf{r}')}{\omega_n^2 + E_\mu^2(\mathbf{k})},$$ (11)

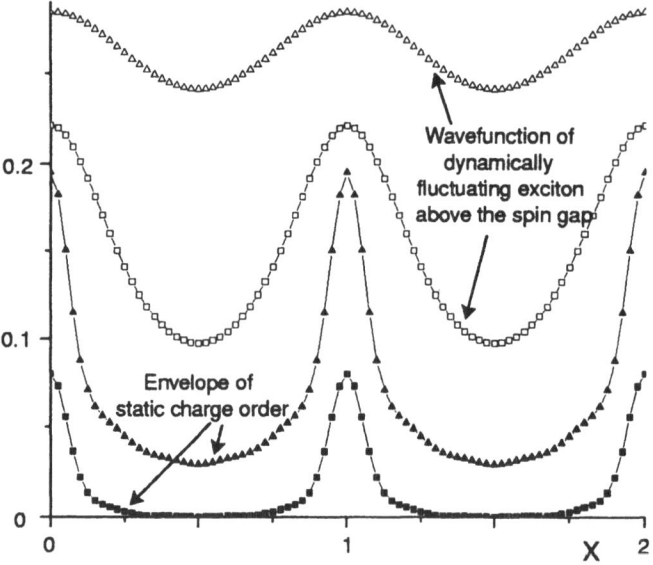

Figure 3.   Spatial structure of the spin and charge correlations in the SC phase at two different magnetic fields (squares and triangles; the squares have the lower field). The vortices are centered at $x = 0, 1, 2$. The orientation of the spins fluctuates with a frequency of order the inverse spin gap $\hbar/\Delta$. The open symbols represent the lowest energy exciton wavefunction $\Xi_{00}(\mathbf{r})$. The nature of the microscopic spin correlations can be understood by recalling that this wavefunction is the envelope of the order Fig 1. After including the pinning term S_{pin}, static charge order is induced, and its envelope $\Omega(\mathbf{r})$ (defined in (12,13)) is shown above. The numerical results were obtained by Ying Zhang and reported in [9].

where $\mathbf{k}$ is a Bloch momentum which extends over the first Brillouin zone of the vortex lattice, μ is a 'band' index, $\Xi_{\mu\mathbf{k}}(\mathbf{r})$ are the Bloch states $(\Xi_{\mu\mathbf{k}}(\mathbf{r}+\mathbf{R}_v) = e^{i\mathbf{k}\cdot\mathbf{R}_v}\Xi_{\mu\mathbf{k}}(\mathbf{r})$ where $\mathbf{R}_v$ is an vector connecting two vortex centers), $E_\mu(\mathbf{k})$ are the energy dispersions of the various bands, and ω_n is an imaginary Matsubara frequency. The lowest energy exciton has energy $E_0(\mathbf{0}) \equiv \Delta$ and wavefunction $\Xi_{00}(\mathbf{r})$; this wavefunction is sketched for typical parameter values in Fig 3.

Now let us consider the influence of the pinning term in (7). Combining (7) with (2) it is simple to see that to first order in the $\zeta_{x,y}$, static charge order appears in the SC phase, with

$$\langle \delta\rho(\mathbf{r}) \rangle \propto \mathrm{Re}\left[\zeta_x e^{i2\mathbf{K}_x\cdot\mathbf{r}} + \zeta_y e^{i2\mathbf{K}_y\cdot\mathbf{r}} \right] \Omega(\mathbf{r}) \qquad (12)$$

where

$$\Omega(\mathbf{r}) \equiv T \sum_{\omega_n} \sum_{\mathbf{r}_v} G_\varphi^2(\mathbf{r}, \mathbf{r}_v, \omega_n). \qquad (13)$$

A plot of the function $\Omega(\mathbf{r})$ is sketched in Fig 3, along with the corresponding $\Xi_{00}(\mathbf{r})$. At the higher field (triangles), the spin exciton wavefunction $\Xi_{00}(\mathbf{r})$ is essentially constant across the entire system, with only a weak modulation induced by the vortex lattice; nevertheless, at the same field the charge order $\Omega(\mathbf{r})$ has a strong modulation on the scale of $c/(2\Delta)$, where c is a spin-wave velocity [8]. At the lower field (squares), there is larger modulation in the spin exciton (but $\Xi_{00}(\mathbf{r})$ only decays to half its maximum value), and again the decay length of the charge correlations is about half that of the spin correlations. The spatial form of the lower field $\Omega(\mathbf{r})$ in Fig 3 is quite similar to envelope of the modulation observed in [1].

The STM experiments of [1] actually measure the modulation in the local electronic density of states (LDOS) in a range of energies as a function position in vortex lattice. A great deal of information is, in principle, contained in the energy and spatial dependence of the LDOS modulations. We have recently analyzed [8, 10] simple models for the coupling of the electronic quasiparticles to the collective SDW and SC degrees of freedom that have been discussed here: these lead to predictions for the LDOS modulations, which can be usefully compared with the STM data—the reader is referred to the papers for details.

3.2. STATIC SPIN MOMENT IN THE SC+SDW PHASE

The SC+SDW phase has $\langle\varphi_\mu\rangle \neq 0$, and hence via (1), (2), and (3), there is both static spin and charge order. Neutron scattering measurements have so far only succeeded in observing the static spin order, and so we will restrict our discussion here to the spin moment.

In the presence of an applied magnetic field, the vortex lattice will spatially modulate the value of $\langle\phi_\mu(\mathbf{r})\rangle$, and this should, in principle, lead to satellite elastic peaks [19, 5, 9] surrounding the main elastic Bragg peaks at $\pm\mathbf{K}_x$, $\pm\mathbf{K}_y$ observed in neutron scattering. However, our discussion above on the dominance of the superflow effect shows that this modulation occurs predominantly on the scale of the vortex lattice spacing [5, 9], and not on the scale of the vortex core [19]. The resulting satellite peaks are then found to be very weak, as will be clear from our results below.

We show a sketch of the spatial form of $\langle\varphi_\mu(\mathbf{r})\rangle$ at a point in the SC+SDW phase in Fig 4. All the coupling constants in the theory are the same as those used for the results in Fig 3 for the SC phase. Only the parameters s and H are tuned to move the system between the SC and SC+SDW phases. The spatial Fourier transform of Fig 4 determines

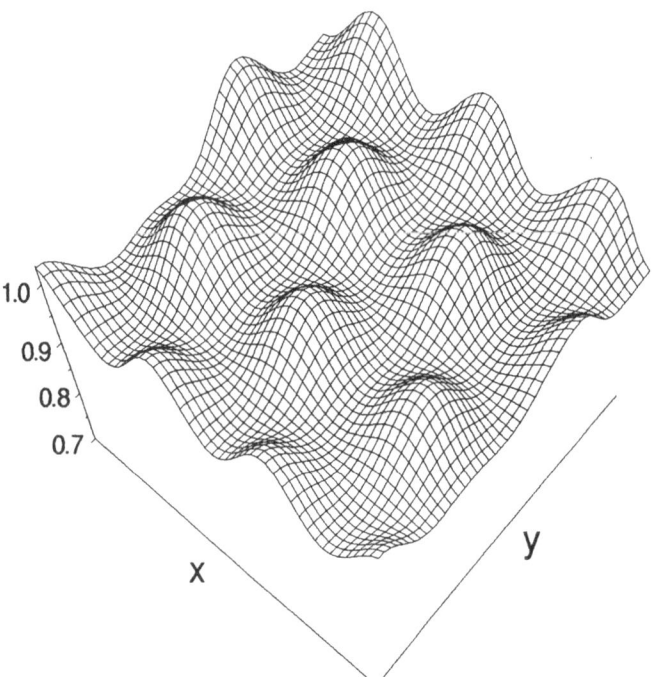

Figure 4.   Spatial form of $\langle \varphi_\mu(\mathbf{r})$ in the SC+SDW phase. Notice the vertical scale, which shows that the overall modulation in the size of the order parameter is quite small. The numerical results were obtained by Ying Zhang and reported in [9].

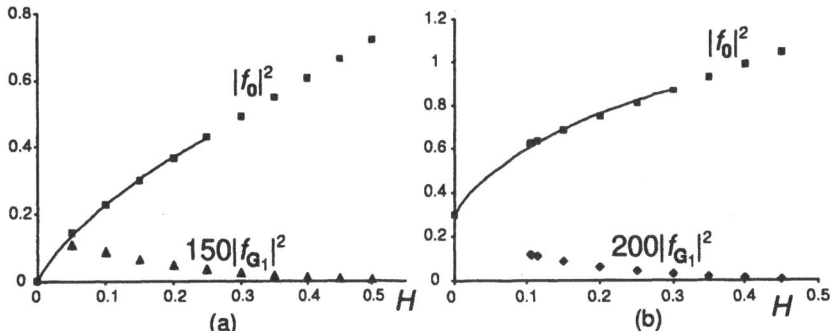

Figure 5. Magnitudes of the elastic scattering peaks in (14) obtained by the spatial Fourier transform of results like those in Fig 4. The field H is measured in units of H_{c2}^0, the value of the critical field at which all four phases meet in Fig 2. The plot (a) is at $s = s_c$, while (b) is for $s < s_c$. All other couplings are identical to those in Figs 3 and 4. The numerical results were obtained by Ying Zhang and reported in [9].

the strength of the elastic Bragg peaks that will be observed in neutron scattering. In particular, the dynamic structure factor of φ_μ has the form

$$S_\varphi(\mathbf{k}, \omega) = (2\pi)\delta(\omega) \sum_{\mathbf{G}} |f_\mathbf{G}|^2 (2\pi)^2 \delta(\mathbf{k} - \mathbf{G}) \tag{14}$$

where $\mathbf{G}$ are the reciprocal lattice vectors of the vortex lattice. The reader should keep in mind, via (1) and (3), that the experimental structure factor is obtained from (14) by measuring wavevectors from the SDW ordering wavevectors $\pm\mathbf{K}_x$, $\pm\mathbf{K}_y$. We show typical results for the field dependence of the strength of the central peak, $|f_0|^2$, and also for the first satellite peak, $|f_{\mathbf{G}_1}|^2$ (where $\mathbf{G}_1$ is the smallest non–zero reciprocal lattice vector) in Fig 5. The strong observable effect is in the H dependence of the central peak $|f_0|^2$. The same superflow effects which were responsible for (10), also dominate in determining the average magnitude of the SDW order in the SC+SDW phase: the enhancement of magnetic order by the superflow leads to the dependence

$$|f_0|^2 \propto H \ln(1/H); \tag{15}$$

the lines in Fig 5 are fits of the full numerical solution to (15). Arovas *et al.* [19] had discussed static magnetic order in the vortex core: the average moment in their model is proportional to the number of vortices, and this leads to a weaker linear H dependence in the average moment. The contribution of the superflow, leading to (15), always dominates in the SC+SDW phase.

The results in Fig 5 also compare well with the experimental observations in the overall scale of both the field and the magnetic moment: those in (a) are quite similar to the results of [2], while (b) matches well with [3].

Finally, note that Fig 5 shows that the satellite peaks are unobservable small. This is related to relatively slow modulation induced by the superflow in the moment in Fig 4. Ref. [9] predicted that the influence of the vortex lattice may be more easily observable in the *dynamic* exciton band structure in the SC phase.

4.   WHY DOES THE CHARGE ORDER HAVE PERIOD 4?

The SDW ($K_{x,y}$) and CDW ($2K_{x,y}$) ordering wavevectors have so far been arbitrary parameters in our phenomenological theory, and the structure of this theory is largely independent of the values of $K_{x,y}$ (some high order terms in the action are permitted only for certain commensurate values of $K_{x,y}$). The determination of $K_{x,y}$ requires, instead, a lattice scale theory of the doped antiferromagnet.

The mechanism of the charge–ordering instability in doped antiferromagnets has been discussed using a number of different theoretical perspectives [22, 23, 24, 25, 26]. In general, several possible charge–ordering periods emerge in these works, and there appear to be no fundamental principles restricting the possible periods, or whether the charge–ordering is site or bond centered. Here, we will briefly recall our theoretical work on charge ordering [4]: its point of departure is the theory of the magnetic quantum critical point in the Mott insulator. Our perspective led to bond–centered charge ordered states, without long-range magnetic order, and with co-existing d–wave–like superconductivity. The predicted evolution of the wavevector of this ordering, as a function of hole concentration, is shown in Fig 6. Note that the period, p, is always pinned to be an even number *i.e.* $K_x = (2\pi/a)(1/p, 0)$, where a is the square lattice spacing. There is a large range of δ values in Fig 6 where the period is pinned at $p = 4$: this corresponds to the values of $K_{x,y}$ observed in the STM experiments [1]. Further support for Fig 6 has emerged from recent neutron scattering measurements of Mook *et al.* [27] in the under–doped superconductor $YBa_2Cu_3O_{6.35}$: they observed static charge order, and dynamic spin correlations, with the the charge order period pinned rather precisely at $p = 8$. This is in accord with the plateau at $p = 8$ over a range of small δ values in Fig 6, and should be contrasted with the continuous evolution of ordering wavevector with

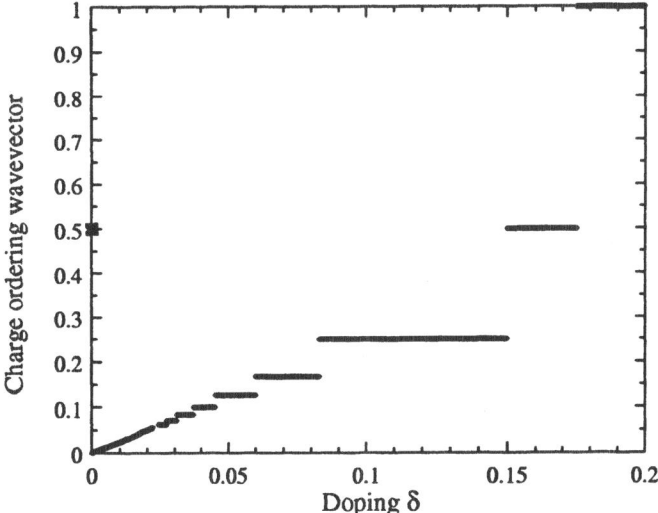

Figure 6. Evolution of the charge ordering wavevector upon doping a paramagnetic Mott insulator, from Ref. [4]. The wavevector is measured in units of $2\pi/a$. At $\delta = 0$, the Mott insulator has period 2, which is associated with the appearance of bond order (see Fig 7). Beyond a range of very small δ values, the ground state is also a superconductor. Full square lattice symmetry is restored above $\delta \approx 0.175$, when the ground state becomes an ordinary d–wave superconductor.

δ which is usually assumed in the "Yamada plot" [28, 29] for periods larger than $p = 4$.

We conclude this paper by briefly recalling the physical ingredients behind the results in Fig 6. A review of these arguments has already been presented in [7], and we present here a synopsis in Fig 7. In moving from an undoped Mott insulator, like La_2CuO_4, to a high temperature superconductor at optimal doping, at least two quantum phase transitions must take place: one involving the loss of magnetic order, and the other the onset of superconductivity. Theoretically, it is very useful to disentangle the two transitions by imagining that we have a second tuning parameter at our disposal, in addition to the doping δ. We use this second parameter to first destroy the magnetic order in La_2CuO_4 while remaining at $\delta = 0$—a specific possibility for such a parameter is a frustrating second neighbor exchange interaction (see Fig 7). Detailed arguments have been given (for a recent review see [30]) that the paramagnetic Mott insulator so obtained has bond-centered charge order, as indicated in Figs 6 and 7. The second theoretical step of doping the paramagnetic Mott insulator can be reliably addressed in a large N theory [4], and one eventually obtains a d–wave superconductor which

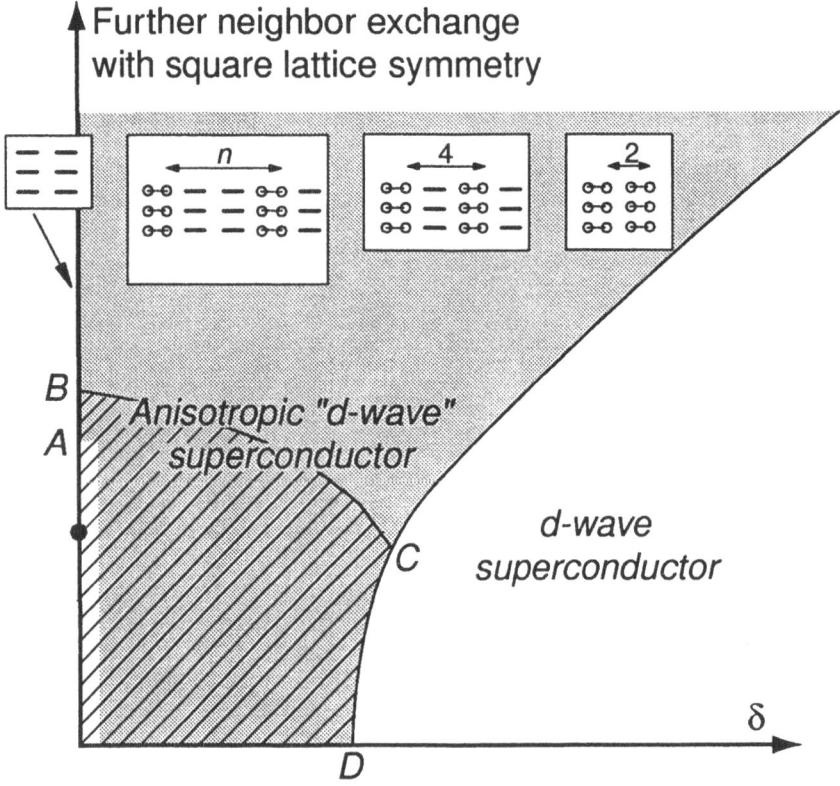

Figure 7. Schematic $T = 0$ phase diagram (from Refs. [4, 7]) for the high temperature superconductors as a function of a ratio of the near neighbor exchange interactions and the hole concentration, δ; *e.g.* the vertical axis could be J_2/J_1, the ratio of the first to second neighbor exchange. The shaded region has charge order. The hatched region has broken spin–rotation symmetry with $\langle \vec{S} \rangle \neq 0$, and at least one of the order parameters $\Phi_{x,y\alpha}$ is non-zero; $\langle \vec{S} \rangle = 0$ elsewhere. The unfrustrated, insulating antiferromagnet with long–range Néel order is indicated by the filled circle. At $\delta = 0$, there is an onset of charge order above the point A, while spin–rotation invariance is restored above B. The nature of the charge orders as determined by the computations of Ref. [4] are indicated at the top of the figure; numerous transitions, within the gray shaded region, in the nature of the charge ordering are not shown. The ground state at very low non-zero doping is an insulating Wigner crystal and there is subsequently a insulator–to–superconductor transition; superconductivity is present over the bulk of the $\delta > 0$ region. The central idea behind our approach is that many essential aspects the spin excitation spectrum of the insulating, paramagnetic region ($\delta = 0$, $\langle \vec{S} \rangle = 0$) lead to a simple and natural description of the analogous properties of the d–wave superconductor.

fully respects the symmetries of spin rotations and lattice translations and rotations. In this approach, the charge order of the superconductor without magnetic long range order evolves from that in the paramagnetic Mott insulator. Further, the Cooper pairing in the superconductor is also connected to the singlet electron pairing present in the bond–ordered paramagnetic Mott insulator. These connections led us to the results in Fig 6, and to the proposal of charge order nucleation by vortices in [6].

Acknowledgments

The results reviewed here were obtained with Eugene Demler, Kwon Park, Anatoli Polkovnikov, Matthias Vojta, and Ying Zhang; I thank them for fruitful collaborations. This research was supported by US NSF Grant DMR 0098226.

References

[1] J.E. Hoffman, E.W. Hudson, K.M. Lang, V. Madhavan, S.H. Pan, H. Eisaki, S. Uchida, and J.C. Davis, Science **295** (2002) 466.

[2] B. Lake, H.M. Rønnow, N.B. Christensen, G. Aeppli, K. Lefmann, D.F. McMorrow, P. Vorderwisch, P. Smeibidl, N. Mangkorntong, T. Sasagawa, M. Nohara, H. Takagi, T.E. Mason, Nature **415** (2002) 299.

[3] B. Khaykovich, Y.S. Lee, S. Wakimoto, K.J. Thomas, R. Erwin, S.-H. Lee, M.A. Kastner, and R.J. Birgeneau, cond–mat/0112505.

[4] M. Vojta and S. Sachdev, Phys. Rev. Lett. **83** (1999) 3916; M. Vojta, Y. Zhang and S. Sachdev, Phys. Rev. **B62** (2000) 6721; S. Sachdev and N. Read, Int. J. Mod. Phys. **B5** (1991) 219.

[5] E. Demler, S. Sachdev, and Y. Zhang, Phys. Rev. Lett. **87** (2001) 067202.

[6] K. Park and S. Sachdev, Phys. Rev. **B64** (2001) 184510.

[7] S. Sachdev, Proceedings of Spectroscopies in Novel Superconductors, J. Phys. Chem. Solids, to appear, cond–mat/0108238.

[8] A. Polkovnikov, S. Sachdev, M. Vojta, and E. Demler, Proceedings of PPHMF IV, World Scientific, Singapore, in press, cond–mat/0110329.

[9] Y. Zhang, E. Demler, and S. Sachdev, cond–mat/0112343.

[10] A. Polkovnikov, M. Vojta, and S. Sachdev, to appear on the cond–mat archive in February/March 2002.

[11] S. Sachdev and J. Ye, Phys. Rev. Lett. **69** (1992) 2411; A.V. Chubukov and S. Sachdev, Phys. Rev. Lett. **71** (1993) 169.

[12] T. Imai, C.P. Slichter, K. Yoshimura, and K. Kosuge, Phys. Rev. Lett. **70** (1993) 1002.

[13] G. Aeppli, T.E. Mason, S.M. Hayden, H.A. Mook, and J. Kulda, Science **278** (1997) 1432.

[14] L. Balents, M.P.A. Fisher, and C. Nayak, Int. J. Mod. Phys. **B12** (1998) 1033.

[15] A.V. Chubukov and S. Sachdev and J. Ye, Phys. Rev. **B49** (1994) 11919; S. Sachdev and M. Vojta, Physica **B280** (2000) 333.

[16] S. Sachdev, C. Buragohain, and M. Vojta, Science **286** (1999) 2479; M. Vojta, C. Buragohain and S. Sachdev, Phys. Rev. **B61** (2000) 15152.

[17] O. Zachar, S.A. Kivelson, and V.J. Emery, Phys. Rev. **B57** (1998) 1422.

[18] S.-C. Zhang, Science **275** (1997) 1089.

[19] D.P. Arovas, A.J. Berlinsky, C. Kallin, and S.-C. Zhang, Phys. Rev. Lett. **79** (1997) 2871; J.-P. Hu and S.-C. Zhang, cond–mat/0108273.

[20] B. Lake, G. Aeppli, K.N. Clausen, D.F. McMorrow, K. Lefmann, N.E. Hussey, N. Mangkorntong, M. Nohara, H. Takagi, T.E. Mason, and A. Schröder, Science **291** (2001) 1759.

[21] S. Sachdev, Phys. Rev. **B45** (1992) 389; N. Nagaosa and P.A. Lee, Phys. Rev. **B45** (1992) 966.

[22] S.R. White and D.J. Scalapino, Phys. Rev. Lett. **80** (1998) 1272; Phys. Rev. **B61** (2000) 6320.

[23] M. Bosch, W.v. Saarloos, and J. Zaanen, Phys. Rev. **B63** (2001) 092501; J. Zaanen and A.M. Oles, Ann. Phys. (Leipzig) **5** (1996) 224.

[24] S.A. Kivelson, E. Fradkin, and V.J. Emery, Nature **393** (1998) 550; U. Low, V.J. Emery, K. Fabricius, and S.A. Kivelson, Phys. Rev. Lett. **72** (1994) 1918.

[25] H. Tsunetsugu, M. Troyer, and T.M. Rice, Phys. Rev. **B51** (1995) 16456.

[26] C. Nayak and F. Wilczek, Phys. Rev. Lett. **78** (1997) 2465.

[27] H.A. Mook, P. Dai, and F. Dogan, Phys. Rev. Lett. **88** (2002) 097004.

[28] J.M. Tranquada, J.D. Axe, N. Ichikawa, A.R. Moodenbaugh, Y. Nakamura, and S. Uchida, Phys. Rev. Lett. **78** (1997) 338.

[29] K. Yamada *et al.*, Phys. Rev. **B57** (1998) 6165.

[30] S. Sachdev and K. Park, Annals of Physics, May 2002, cond–mat/0108214.

THERMODYNAMIC ANALOGIES BETWEEN BOSE–EINSTEIN CONDENSATION AND BLACK-BODY RADIATION

Victor Romero–Rochin

Instituto de Física, Universidad Nacional Autónoma de México.
Apartado Postal 20-364, 01000 México, D.F, México.
romero@fisica.unam.mx

Vanderlei S. Bagnato

Instituto de Fisica, Universidade de São Paulo.
C. Postal 369, 13560-970 São Carlos, SP, Brasil.
vander@if.sc.usp.br

Abstract   We discuss the thermodynamic properties of a bosonic gas trapped in a harmonic potential. We identify properly the extensive thermodynamic variable equivalent to the volume and the intensive thermodynamic variable equivalent to the pressure. With these variables one is able to deduce all the thermodynamics of the system. From this point of view, an interesting analogy with Black–Body radiation emerges, showing that at and below the critical temperature, the non–condensate fraction of atoms behaves thermodynamically like a gas of massless particles.

Keywords: Bose–Einstein Condensation, Black–Body Radiation.

1.   INTRODUCTION

The achievement of Bose–Einstein condensation (BEC) in dilute atomic gases has open a new window into the quantum world and offers profound insights into macroscopic manifestations of quantum phenomena. [1, 2, 3]

An important feature associated with available experimental traps of atoms is that the confining potential can be safely approximated by a harmonic–type dependence. Thus, most of the recent theoretical analysis also deal with BEC under those conditions. [4, 5, 6] The main con-

Developments in Mathematical and Experimental Physics,
Volume B: Statistical Physics and Beyond
Edited by Macias *et al.*, Kluwer Academic/Plenum Publishers, 2003

187

cern of current research has been the study of the condensate fraction of atoms [7, 8, 9] and little attention has been paid to the non–condensate fraction present at finite temperatures. In the case of finite temperature condensates we have the situation where only part of the atomic cloud is participating in the macroscopic occupation of the ground state. The properties of the latter have been the subject of current theoretical and experimental scrutiny and all the extracted properties from the mixed cloud of atoms are mainly obtained from the analysis of the images containing the density profiles. [10]

An aspect that has not been fully explored concerns the thermodynamics of a BEC in a harmonic trap. Identification of *extensive* and *intensive* macroscopic variables (like volume and pressure respectively) and their interconnection in an equation of state has not been done yet. The identification of such variables allows an easy analysis for a large variety of properties including heat capacities, and other susceptibilities, and a more traditional view of BEC in a gas trapped by an inhomogeneous potential in terms of isothermal curves.

In this article we consider an ideal bosonic gas trapped by a harmonic potential. Extensive and intensive variables are properly identified allowing for a full analysis of the thermodynamics of the system. Investigating the properties of the non–condensate fraction at and below T_c (critical temperature) an intriguing behavior is obtained: the non–condensate cloud obeys the same thermodynamics of the well known Black–Body Radiation (BBR) problem. In other words, the non–condensate fraction of the harmonic trapped cloud behaves thermodynamically as a gas of massless particles. This coincidence may reveal a more fundamental relation between BEC and BBR which certainly demands more consideration and experimental exploration.

2.   THERMODYNAMICS OF A GAS OF BOSONS IN A HARMONIC TRAP

We start by considering N bosons of mass m in a 3D isotropic harmonic trap, characterized by the Hamiltonian

$$H = \sum_{i=1}^{N} \hbar\omega(\hat{n}_i^x + \hat{n}_i^y + \hat{n}_i^z) \tag{1}$$

with the zero–point energy equal to zero and where $\hat{n}_i$ are the number operators.

The Grand Potential in the Grand Canonical ensemble is given by ($\alpha = \mu/kT$) [11]

$$\Omega(\mu, T, \omega) = kT \sum_{ijk} \ln\left(1 - e^{\beta\hbar\omega(i+j+k)+\alpha}\right) \qquad (2)$$

where i, j, k are integers that run from 0 to ∞. For the 3D harmonic potential the density of states is $\rho(L) = L^2/2$, where $L = i+j+k$ is the dimensionless energy of one particle. The Grand Potential can then be written as

$$\Omega(\mu, T, \omega) = -\frac{kT}{6}\frac{1}{(\beta\hbar\omega)^3}\int_0^\infty dx \frac{x^3}{e^{x-\alpha} - 1} \qquad (3)$$

where, as usual, the zero–point energy contribution arising from $i = j = k = 0$ has been excluded.

Introducing the Bose function,

$$g_n(\alpha) = \frac{1}{\Gamma(n)}\int_0^\infty dx \frac{x^{n-1}}{e^{x-\alpha} - 1} \qquad (4)$$

we can therefore write

$$\Omega(\mu, T, \omega) = -kT\left(\frac{kT}{\hbar\omega}\right)^3 g_4(\alpha). \qquad (5)$$

Following the standard procedure of thermodynamics, the average number of particles N, the average internal energy E and the entropy S, as functions of μ, T, ω, can be obtained: [4]

$$N = \left(\frac{kT}{\hbar\omega}\right)^3 g_3(\alpha), \qquad (6)$$

$$E = 3kT\left(\frac{kT}{\hbar\omega}\right)^3 g_4(\alpha) \qquad (7)$$

and

$$S = k\left(\frac{kT}{\hbar\omega}\right)^3 [4g_4(\alpha) - \alpha g_3(\alpha)]. \qquad (8)$$

Here, by following standard thermodynamics, we notice that the right-hand–side of Ω in Eq.(5) must be equal to (minus) the product of an *intensive* thermodynamic variable times an *extensive* thermodynamic variable. The former being the analog of the pressure while the latter is equivalent to the volume. The physical interpretation of those variable must be done with special care since the inhomogeneous potential does not allow for a simple characterization of pressure and volume. At this point we simple propose that the *extensive* variable be

$$V = \frac{1}{(\hbar\omega)^3} \tag{9}$$

and thus, from the expression for the Grand potential (5), the *intensive* variable must be

$$p = kT(kT)^3 g_4(\alpha). \tag{10}$$

We shall call these quantities the "volume" and the "pressure" respectively. That these quantities do not have the units of volume and pressure is irrelevant. The volume is certainly related to the spatial extension of the system, while pressure with the response of the gas to the external confinement. Their product has the correct units of energy. It is certainly clear that the frequency ω of the trap must play the role of the thermodynamic volume since, after all, it is the external field. What it is not so obvious is that it is ω^{-3} the proper quantity. There are several ways in which one can check that this is the correct interpretation. First, it is clear from Eqs.(6)-(8) and (9) that if $V \to \lambda V$ then $N \to \lambda N$, $E \to \lambda E$, and $S \to \lambda S$; that is, it obeys the definition of extensiveness. Second, the first law of thermodynamics can be written as

$$dE = TdS - pdV + \mu dN. \tag{11}$$

Thus, if we calculate the change in internal energy due to an adiabatic "expansion" or "compression", namely a change of frequency ω at constant N, we find the same result both from the expressions for N, E and S given by Eqs.(6)-(8) and from the first law, Eq.(11). We also find that the adiabatic invariants are $\mu/kT = const$ and $\hbar\omega/kT = const$.

Once we have all thermodynamic variables S, N, V and their conjugates T, μ, p, we can proceed and do the standard thermodynamics.

¿From the above expressions we find that the internal energy and the pressure may be written as

$$E = 3NkT\frac{g_4(\alpha)}{g_3(\alpha)} \quad \text{and} \quad p = \frac{NkT}{V}\frac{g_4(\alpha)}{g_3(\alpha)} \tag{12}$$

Combination of these two yields $E = 3pV$. For high temperatures all the Bose functions become equal, $g_n(\alpha) \approx \exp(\alpha)$ and the equations of state approach those of a classical gas, *but* of massless or ultrarelativistic particles.

¿From the expressions for the pressure (12) and the number of particles (6) we can calculate the isotherms $p - -V$ at constant number of particles. This is shown in Fig. 1. The qualitative structure of the

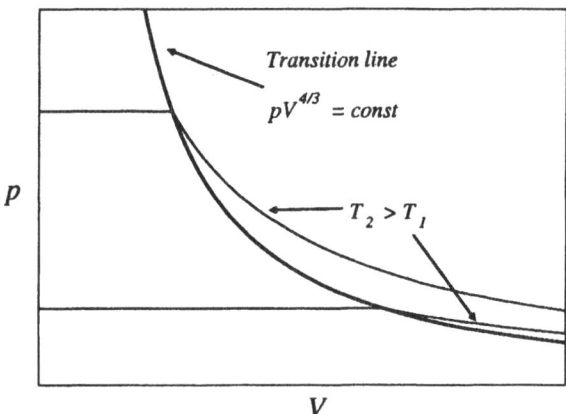

Figure 1. Isotherms of the $p - -V$ equation of state at constant N, cf. Eq.(12). See text for proper definition of volume and pressure.

isotherms is equal to the usual case of free bosons discussed in many textbook [11]. For large volumes (i.e. low densities), the isotherms approach those of an ideal classical gas; as the volume is lowered Bose–Einstein condensation sets in at a critical value given by the curve $pV^{4/3} = const$ as shown in Fig. 1. Below the critical value, at constant temperature, only the non–condensate fraction of atoms exert pressure although this remains constant all the way to zero volume; the excess number of particles condensate into the $i = j = k = 0$ ground state. The main difference of our analysis with respect to the free–BEC case is that the in the latter the critical line scales as $pV^{5/3} = const$.

With the identification of volume and pressure we can now calculate the heat capacities at constant volume and pressure, C_V and C_p; the isothermal compressibility at constant number of particles, κ_T; and the coefficient of thermal expansion α_T. Here, we report the heat capacities only. These are:

$$C_V = 3Nk \left[4\frac{g_4(\alpha)}{g_3(\alpha)} - 3\frac{g_3(\alpha)}{g_2(\alpha)} \right] \quad \text{for} \ \ T \geq T_c^+ \tag{13}$$

and

$$C_V = 12kV \left(kT \right)^3 g_4(0) \quad \text{for} \ \ T \leq T_c^-, \tag{14}$$

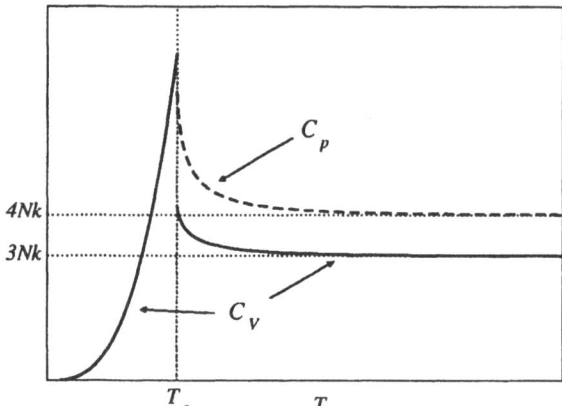

Figure 2.   Heat capacities at constant volume C_V and at constant pressure C_p, as functions of temperature T. Note the discontinuity at T_c and the classical equipartition values $C_V \to 3NK$ and $C_p \to 4Nk$ at high temperature. The behavior of C_V for $T \leq T_c$ is *identical* to that of a photon gas in thermal equilibrium. Below T_c, C_p diverges since pressure only depends on temperature.

The discontinuity at T_c, $C_V(T_c^-) - C_V(T_c^+)$, agrees with the previous calculation of Ref. [4]. See Fig.2. Similarly,

$$C_p = 4Nk\frac{g_4(\alpha)g_2(\alpha)}{g_3^2(\alpha)}\left[4\frac{g_4(\alpha)}{g_3(\alpha)} - 3\frac{g_3(\alpha)}{g_2(\alpha)}\right] \text{ for } T \geq T_c^+. \quad (15)$$

For $T \leq T_c^-$, one finds that $C_p = \infty$ (the same is found for κ_T and α_T). This is not surprising since below T_c the pressure *only* depends on temperature; that is, if the pressure is kept constant, the temperature cannot be raised no matter how much heat one puts into the system. Note that the ratio $C_p/C_V \to 4/3$ for high temperatures; again, this result is the same of an ultrarelativistic gas. Fig. 2 also shows $C_p(T)$.

3.   ANALOGY OF A GAS OF PHOTONS WITH THE NON–CONDENSATE FRACTION OF ATOMS

Here, we briefly explore the very close and intriguing analogy between the Black–Body radiation (BBR) problem and the BEC case in a harmonic potential. We find that, at and below T_c, the gas of massive bosons in the non–condensate fraction behaves thermodynamically *identically* to a gas of photons in thermal equilibrium inside a cavity.

The Grand Potential in the BBR problem can be cast as, [11]

$$\Omega_{BB}(T, V) = -16\pi kTV \left(\frac{kT}{hc}\right)^3 g_4(0) \tag{16}$$

where V is the actual volume of the cavity. Compare this expression with equation (5) for Ω of the gas of bosons in the harmonic trap. The latter expression, by *arbitrarily* multiplying and dividing by c^3, becomes

$$\Omega(\mu, T, \omega) = -kT \left(\frac{2\pi c}{\omega}\right)^3 \left(\frac{kT}{hc}\right)^3 g_4(\alpha). \tag{17}$$

The quantity $(2\pi c/\omega)^3$ now does have units of volume and, by comparing Eq.(17) with Eq.(16) of BBR, we see that indeed it takes the role of the volume. The rest of the terms in Eq.(17) now have units of pressure and its dependence with T is that of a radiation pressure. Since at and below the BEC transition the chemical potential becomes zero, $\alpha = 0$, it follows from the above two equations that the thermodynamics of the photon gas and of the non–condensate fraction of atoms are identical. In the previous section we saw the thermodynamic properties in the classical limit are those of a gas of ultrarelativistic particles.

An important reason for the analogy lies in the fact that zero chemical potential means that the number of particles is no longer a conserved thermodynamic variable. However, this also happens in the usual free–BEC problem without any relationship with BBR. The other reason is the fact that the bosons in the harmonic trap as well as the photons in a cavity are harmonic oscillators (and expressed as such in a second-quantization scheme). These two points together may suffice to account for the analogy for temperatures at and below the critical temperature.

It is of interest to point out that in the present problem, *cf.* Eqs.(5), (6)-(8), the thermal de Broglie wavelength does not appear naturally. The thermodynamic quantity that determines whether the system is in the quantum or in the classical regime is now, see Eq.(6), $N(\hbar\omega/kT)^3$. i.e. if this quantity is of the order of one or less, the system is in the quantum regime. One sees that what determines the crossover from classical to quantum regimes is the ratio of the "volume" per particle $((1/\omega)^3/N)$ to the cube of a thermal quantum "wavelength" (h/kT). The latter, a thermal photon–like wavelength, supersedes the role of the de Broglie wavelength. Notice also that the mass dependence only appears through the frequency ω of the trap.

4. CONCLUSIONS

We have provided a simple approach to evaluate the thermodynamics for a confined boson gas in a harmonic trap through the identification of an extensive variable ("volume") and an intensive variable ("pressure"). Equations of state as well as susceptibilities and heat capacities can be easily evaluated. We found that at and below T_c the remaining atoms not in the condensate behave like a massless particle gas (photons) in perfect analogy with the thermodynamics of Black–Body radiation. One may think that the non-condensate fraction is a photon–like gas that at T_c its particles acquire mass (i.e. when the chemical potential becomes non-zero). This analogy may carry with it more fundamental consequences.

Acknowledgments

Work partially supported by Fapesp (Center for Optics and Photonic Research) and CNPq (Brazil) and by CONACyT (Mexico) through grant 32634–E.

References

[1] M.R. Andrews, C.G. Townsend, H.J. Miesner, D.S. Dustee, D.M. Kurn, and W. Ketterle, Science **275** (1997) 637.

[2] A.J. Legget, Phys. Scripta **T76** (1998) 199.

[3] K. Burnett, M. Edwards, and C. Clark, Phys. Today **52** (1999) 37.

[4] V.S. Bagnato, D.E. Pritchard, and D. Kleppner, Phys. Rev. **A35** (1987) 4354.

[5] J. Schneider and H. Wallis, Phys. Rev. **A57** (1998) 1253.

[6] F. Daltovo, S. Giorgini, L.P. Pitaevskii, and S. Stringari, Rev. Mod. Phys. **71** (1999) 463.

[7] D.M. Stamper, H.J. Miesner, S. Inowye, M. Andrews, and W. Ketterle, Phys. Rev. Lett. **81** (1998) 500.

[8] K.P. Martin and W. Zhang, Phys. Rev. **A57** (1998) 4761.

[9] W. Ketterle and H.J. Miesner, Phys. Rev. **A56** (1997) 3291.

[10] L.U. Hau, B.B. Busch, C. Liu, Z. Dutton, M.M. Burns, and J.A. Golovchenko, Phys. Rev. **A58** (1998) R54.

[11] R.K. Pathria, Statistical Mechanics, (Pergamon Press, Oxford, 1972).

[12] H.B. Callen, Thermodynamics (J. Wiley and Sons, New York, 1960).

[13] M. Anderson, J. R. Ensher, M.R. Mathews, C. E. Weiman, and E. A. Cornell, Science **269** (1995) 198.

IV

LIQUID CRYSTALS AND MOLECULAR FLUIDS

NONEQUILIBRIUM THERMAL LIGHT SCATTERING FROM NEMATIC LIQUID CRYSTALS

R. F. Rodriguez

Instituto de Física,
Universidad Nacional Autónoma de México,
Apdo. Postal 20-364, 01000 México, D.F., México
zepeda@fisica.unam.mx

J. F. Camacho

Instituto de Física,
Universidad Nacional Autónoma de México,
Apdo. Postal 20-364, 01000 México, D.F., México

Abstract   The effects of orientational fluctuations induced by an external temperature gradient, on the dynamic structure factor of a nematic liquid crystal confined in a homeotropic cell are investigated. Using a fluctuating hydrodynamics approach we calculate the equilibrium and nonequilibrium orientational fluctuations correlation functions associated with the longitudinal modes of the system. Then their influence on the light scattering spectrum is analyzed. The equilibrium part of the spectrum is calculated exactly, while the nonequilibrium contribution is evaluated approximately. We find that the intensity of the central peak increases and depends on the relative orientations of the temperature gradient, director and wavevector. Its maximum shifts towards positive values of the frequency shift. The intensities of the sound propagation modes are unequal and one of the peaks shrinks in the same amount as the other increases. Near equilibrium these changes depend linearly on the magnitude of the dimensionless applied gradient which in our model has the value 0.5 for a plate separation of $\sim 10^{-2}$ *cm.*

Keywords: Liquid crystals, Nonequilibrium, Light scattering

Developments in Mathematical and Experimental Physics,
Volume B: Statistical Physics and Beyond
Edited by Macias *et al.*, Kluwer Academic/Plenum Publishers, 2003

1.   INTRODUCTION

The study of hydrodynamic fluctuations about nonequilibrium stationary states of simple fluids attracted a good deal of attention in the past. In particular, the case where the nonequilibrium state was induced by a constant temperature gradient was investigated by using a variety of theoretical approaches. Induced asymmetries in the Brillouin components of the light scattering spectrum were predicted and the long ranged character of the temperature and density fluctuations was established [1], [2], [3], [4], [5], [6], [7]. Some of these predictions have also been observed experimentally [8], [10], [11].

However, in spite of the fact that for liquid crystals the scattered light intensity is several orders of magnitude larger than for ordinary fluids [12], theoretical and experimental studies of nonequilibrium hydrodynamic fluctuations in liquid crystals are rather scarce. On the theoretical side and for the transverse modes only, the effect of orientational fluctuations on the dynamic structure factor of a thermotropic nematic in stationary states generated by a static temperature gradient [13], a stationary shear flow [14] and by a constant pressure gradient [15] have been considered. In the first two cases it was found that the effects of nonequilibrium fluctuations on the dynamic structure factor are rather small, although in principle detectable. But in the case where a constant pressure gradient produces an incompressible planar Poiseuille flow, there occur large changes in the nonequilibrium part of the spectrum which might be observable. The main purpose of this work is to investigate the effect produced by orientational fluctuations associated with the longitudinal modes of a nematic layer confined between two rigid plates and maintained in a stationary steady state by the action of an external temperature gradient. The dynamics of these fluctuations is more complicated because in contrast to the transverse modes case, they are coupled with density, energy and velocity fluctuations. More specifically, we calculate the equilibrium and nonequilibrium parts of the induced dynamic structure factor. The former contribution is calculated exactly, while the second one is only considered in an approximated form. Our results present the trend of change of the spectrum associated with longitudinal modes due to the presence of an external temperature gradient and are the analog of the same problem studied experimentally by Beysens et al [9] for simple fluids. However, to our knowledge, these effects have not been studied experimentally for liquid crystals.

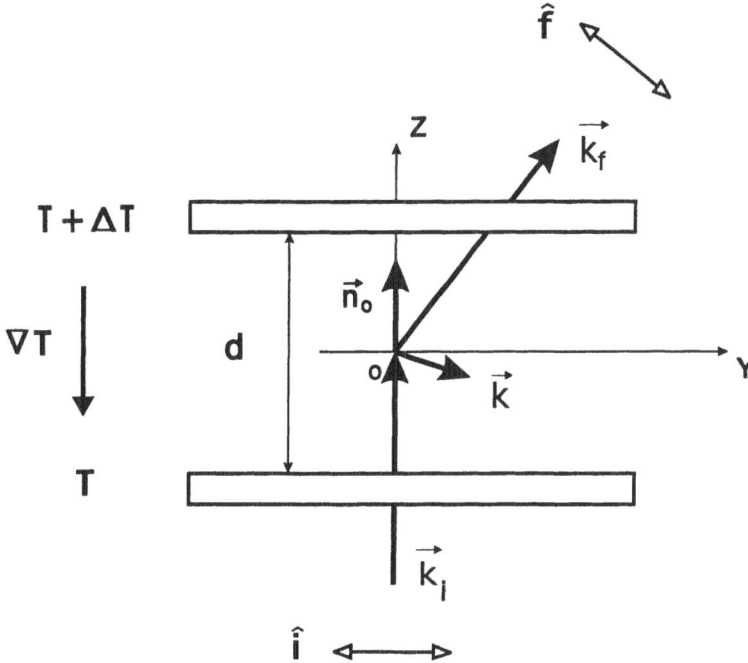

Figure 1. Schematic representation of a homeotropic nematic cell with a constant temperature gradient applied in the $\hat{z}$ direction.

2. MODEL AND BASIC EQUATIONS

Consider a planar homeotropic nematic cell of thickness d with a constant external temperature gradient applied between its plates, as depicted in Figure 1. The gradient is of the form

$$\vec{\nabla}T = - \mid \vec{\nabla}T \mid \hat{z} \equiv -\alpha\hat{z}, \tag{1}$$

where $\hat{z}$ denotes the unit vector along the z direction.

For this arrangement, the chosen initial director's orientation $\hat{n}_0$ along the z axis, is a preferred direction and the hydrodynamic variables may be divided into two independent sets, namely, transverse and longitudinal to $\hat{n}_0$ and the wave vector $\vec{k}$, also defined in Figure 1, [16]. The former set is $\{n_x(\vec{r},t), v_x(\vec{r},t)\}$, while the latter is $\{\rho(\vec{r},t), \sigma(\vec{r},t), v_y(\vec{r},t), v_z(\vec{r},t), n_y(\vec{r},t)\}$. Here $\rho(\vec{r},t)$ and $\sigma(\vec{r},t)$ are, respectively, the local mass density and the entropy per unit mass; $\hat{n}(\vec{r},t)$ is the unitary director field, while $\vec{v}(\vec{r},t)$ denotes the hydrodynamic velocity field. Our purpose in this work is to calculate the contribution to the dynamic structure factor due to these longitudinal variables when the system is in a steady state induced by the constant thermal gradient (1).

The nonequilibrium state to be considered is then characterized by the linear temperature profile

$$T(z) = T_0 - z\alpha, \tag{2}$$

and the induced thermal fluctuations from an equilibrium reference state, identified by a subindex 0, are defined as $\delta\rho(y, z, t) \equiv \rho(y, z, t) - \rho_0$, $\delta\sigma(y, z, t) \equiv \sigma(y, z, t) - \sigma_0$, $\delta v_i(y, z, t) \equiv v_i(y, z, t) - v_{0i}$ ($i = y, z$), $\delta n_y(y, z, t) \equiv n_y(y, z, t) - n_{0y}$. Near equilibrium, the dynamics of these fluctuations is governed by the linearized version of the general nematodynamic equations [17], [18]. By following the well established method of fluctuating hydrodynamics [19], [20], and for the model under consideration, we arrive at the following set linearized fluctuating hydrodynamic equations [21]

$$\frac{\partial}{\partial t}\delta\rho + \rho_0 \frac{\partial}{\partial y}\delta v_y + \rho_0 \frac{\partial}{\partial z}\delta v_z = 0, \tag{3}$$

$$\rho_0 \frac{\partial}{\partial t}\delta v_z - \left(\eta_3 \frac{\partial^2}{\partial y^2} + \eta_1 \frac{\partial^2}{\partial z^2}\right)\delta v_z - (\eta_3 + \eta_5)\frac{\partial^2}{\partial z\partial y}\delta v_y + \frac{c_s^2}{\gamma}\frac{\partial}{\partial z}\delta\rho$$

$$+ \rho_0 c_v \frac{\partial}{\partial z}\delta T + \frac{1}{2}(\lambda - 1)\frac{\partial}{\partial y}\left(K_1 \frac{\partial^2}{\partial y^2} + K_3 \frac{\partial^2}{\partial z^2}\right)\delta n_y$$

$$+ \frac{\partial}{\partial y}\Sigma_{zy} + \frac{\partial}{\partial z}\Sigma_{zz} = 0, \tag{4}$$

$$\rho_0 \frac{\partial}{\partial t}\delta v_y - (\eta_2 + \eta_4)\frac{\partial^2}{\partial y^2}\delta v_y - \eta_3 \frac{\partial^2}{\partial z^2}\delta v_y - (\eta_3 + \eta_5)\frac{\partial^2}{\partial z\partial y}\delta v_z +$$

$$\frac{c_s^2}{\gamma}\frac{\partial}{\partial y}\delta\rho + \rho_0 c_v\theta\frac{\partial}{\partial y}\delta T + \frac{1}{2}(\lambda + 1)\frac{\partial}{\partial z}\left(K_1 \frac{\partial^2}{\partial y^2} + K_3 \frac{\partial^2}{\partial z^2}\right)\delta n_y$$

$$+ \frac{\partial}{\partial y}\Sigma_{yy} + \frac{\partial}{\partial z}\Sigma_{yz} = 0, \tag{5}$$

$$\rho_0 c_v \frac{\partial}{\partial t}\delta T - \left(\kappa_\perp \frac{\partial^2}{\partial y^2} + \kappa_\parallel \frac{\partial^2}{\partial z^2}\right)\delta T + \frac{\rho_0 c_s^2}{\theta}\left(1 - \gamma^{-1}\right)\left(\frac{\partial}{\partial y}\delta v_y + \frac{\partial}{\partial z}\delta v_z\right)$$

$$+ \frac{\partial}{\partial y}g_y + \frac{\partial}{\partial z}g_z = 0, \tag{6}$$

$$\frac{\partial}{\partial t}\delta n_y - \frac{1}{\gamma_1}\left(K_1 \frac{\partial^2}{\partial y^2} + K_3 \frac{\partial^2}{\partial z^2}\right)\delta n_y - \frac{1}{2}(\lambda - 1)\frac{\partial}{\partial y}\delta v_z$$

$$- \frac{1}{2}(\lambda + 1)\frac{\partial}{\partial z}\delta v_y + \Upsilon_y = 0. \tag{7}$$

Here Σ_{zy}, Σ_{zz}, Σ_{yy}, Σ_{yz} are the stochastic components of the linearized currents associated with the stress tensor, g_y, g_z are those of the heat flux vector and Υ_y is the quasi–current in the relaxation of the director. In writing these equations we used the thermodynamic relation

$$\delta T = \frac{T_0 \theta}{\rho_0} \delta \rho + \frac{\rho_0 T_0}{c_v} \delta \sigma, \tag{8}$$

to replace $\delta\sigma(y,z,t)$ by $\delta T(y,z,t)$, with $\theta = \left(\frac{\partial \ln T_0}{\partial \ln \rho_0}\right)_\sigma$; c_v denotes the heat capacity of a unit volume; c_s is the adiabatic sound speed; K_1 and K_3 represent the splay and bend elastic constants; η_i with $i = 1, 2, ...5$, denote the different nematic viscosity coefficients; $\lambda \equiv -\frac{\gamma_2}{\gamma_1}$, is the dimensionless reactive coefficient where γ_1 and γ_2 are parameters with the dimensions of viscosity that determine the relaxation time for the director. $\kappa_\parallel$ and $\kappa_\perp$ describe the thermal conductivity in the directions longitudinal and transverse relative to $\hat{n}$.

Since we are considering a situation near equilibrium, the entropy serves as a potential governing the fluctuations and, therefore, fluctuation-dissipation relations should be valid for the stochastic currents in Eqs. (3) – (7). Following the procedure sketched in the appendix in Ref. [15], [18], we can show that the corresponding fluctuation–dissipation relations have the forms

$$\langle \Sigma_{\alpha j}(\overrightarrow{r}, t) \Sigma_{\beta l}(\overrightarrow{r}', t') \rangle = 2k_B T_0 \nu^0_{\alpha\beta jl} \delta\left(\overrightarrow{r} - \overrightarrow{r}'\right) \delta\left(t - t'\right), \tag{9}$$

$$\langle g_i(\overrightarrow{r}, t) g_j(\overrightarrow{r}', t') \rangle = 2k_B T_0^2 \kappa^0_{ij} \delta\left(\overrightarrow{r} - \overrightarrow{r}'\right) \delta\left(t - t'\right), \tag{10}$$

$$\langle \Upsilon_\mu(\overrightarrow{r}, t) \Upsilon_\nu(\overrightarrow{r}', t') \rangle = 2k_B T_0 \frac{1}{\gamma_1} \delta^{0\perp}_{\mu\nu} \delta\left(\overrightarrow{r} - \overrightarrow{r}'\right) \delta\left(t - t'\right). \tag{11}$$

Here

$$\nu^0_{ijkl} \equiv \nu_2(\delta_{jl}\delta_{ik} + \delta_{il}\delta_{jk}) + 2(\nu_1 + \nu_2 - 2\nu_3)n_{oi}n_{oj}n_{ok}n_{ol}$$
$$+(\nu_3 - \nu_2)(n_{oj}n_{ol}\delta_{ik} + n_{oj}n_{ok}\delta_{il} + n_{oi}n_{ok}\delta_{jl} + n_{oi}n_{ol}\delta_{jk})$$
$$+(\nu_4 - \nu_2)\delta_{ij}\delta_{kl} + (\nu_5 - \nu_4 + \nu_2)(\delta_{ij}n_{ok}n_{ol} + \delta_{kl}n_{oi}n_{oj}), \tag{12}$$

$$\kappa^0_{ij} = \kappa_\perp \delta_{ij} + \kappa_a n_{oi} n_{oj}, \tag{13}$$

where $\kappa_a \equiv \kappa_\parallel - \kappa_\perp$ is the anisotropy in the thermal conductivity, $\delta^{0\perp}_{ij} \equiv \delta_{ij} - n_{oi}n_{oj}$ and k_B denotes Boltzmann's constant.

Defining the Fourier transform of an arbitrary field $\overrightarrow{A}(\overrightarrow{r}, t)$ by

$$\widetilde{\overrightarrow{A}}(\overrightarrow{k}, \omega) \equiv \frac{1}{(2\pi)^4} \int_{-\infty}^{\infty} \overrightarrow{A}(\overrightarrow{r}, t) e^{-i(\overrightarrow{k} \cdot \overrightarrow{r} - \omega t)} d\overrightarrow{r} \, dt \tag{14}$$

with

$$\vec{A}(\vec{r},t) = \int_{-\infty}^{\infty} \widetilde{\vec{A}}(\vec{k},\omega)e^{i(\vec{k}\cdot\vec{r}-\omega t)}d\vec{k}\,d\omega, \tag{15}$$

the Fourier transforms of (3) - (7) may be written in matrix form as

$$M_{ij}u_j = f_i, \tag{16}$$

where M_{ij} is the matrix

$$\begin{pmatrix}
\omega & 0 & -\rho_0 k_y & -\rho_0 k_z & 0 \\
0 & \omega + i\gamma D_T k^2 & -\frac{c_s^2}{c_v\theta}(1-\frac{1}{\gamma})k_y & -\frac{c_s^2}{c_v\theta}(1-\frac{1}{\gamma})k_z & 0 \\
-\frac{c_s^2}{\gamma\rho_0}k_y & -c_v\theta k_y & \omega + iD_\xi k^2 & iD_\mu k^2 & \frac{\eta_1\lambda_1}{\rho_0}D_\gamma k_z k^2 \\
-\frac{c_s^2}{\gamma\rho_0}k_z & -c_v\theta k_z & iD_\mu k^2 & \omega + iD_\nu k^2 & \frac{\eta_1\lambda_0}{\rho_0}D_\gamma k_y k^2 \\
0 & 0 & \lambda_1 k_z & \lambda_0 k_y & \omega + iD_\gamma k^2
\end{pmatrix} \tag{17}$$

$$u_i(k_y,k_z,\omega) = \begin{pmatrix} \delta\widetilde{\rho} \\ \delta\widetilde{T} \\ \delta\widetilde{v}_y \\ \delta\widetilde{v}_z \\ \delta\widetilde{n}_y \end{pmatrix} \tag{18}$$

and

$$f_i(k_y,k_z,\omega) = \frac{1}{\rho_0}\begin{pmatrix} 0 \\ c_v^{-1}(k_y\widetilde{g}_y + k_z\widetilde{g}_z) \\ k_y\widetilde{\Sigma}_{yy} + k_z\widetilde{\Sigma}_{yz} \\ k_y\widetilde{\Sigma}_{zy} + k_z\widetilde{\Sigma}_{zz} \\ -i\rho_0\widetilde{\Upsilon}_y \end{pmatrix}. \tag{19}$$

In the above expressions the following abbreviations have been used, $k^2 = k_y^2 + k_z^2$, $\lambda_0 \equiv \frac{1}{2}(\lambda - 1)$, $\lambda_1 \equiv \frac{1}{2}(\lambda + 1)$, $D_T = \frac{\tau_k}{\rho_0 C_p}$, $D_\xi = \frac{\xi_k}{\rho_0}$, $D_\nu = \frac{\nu_k}{\rho_0}$, $D_\gamma = \frac{K_k}{\eta_1}$, $D_\mu = \frac{\mu_k}{\rho_0}$, with

$$\tau_k \equiv \frac{1}{k^2}\left(\kappa_\perp k_y^2 + \kappa_\parallel k_z^2\right), \xi_k \equiv \frac{1}{k^2}\left[(\eta_2 + \eta_4)k_y^2 + \eta_3 k_z^2\right],$$

$$\nu_k \equiv \frac{1}{k^2}\left(\eta_3 k_y^2 + \eta_1 k_z^2\right), K_k \equiv \frac{1}{k^2}\left(K_1 k_y^2 + K_3 k_z^2\right),$$

$$\mu_k \equiv \frac{1}{k^2}(\eta_3 + \eta_5)k_z k_y, \tag{20}$$

and the following thermodynamic relation was employed

$$\frac{1}{\gamma} = 1 - \frac{C_v\theta^2 T_0}{c_s^2}, \tag{21}$$

being $\gamma = \frac{c_p}{c_v}$ where c_p is the specific heat at constant pressure.

The Fourier transforms of the fluctuation–dissipation relations (9) – (11) turned out to be

$$\left\langle \widetilde{\Sigma}_{\alpha j}(\overrightarrow{k},\omega)\widetilde{\Sigma}_{\beta l}(\overrightarrow{k}',\omega') \right\rangle = 2k_B \nu^0_{\alpha j \beta l} \widetilde{T}\left(\overrightarrow{k} - \overrightarrow{k}'\right) \delta\left(\omega - \omega'\right), \quad (22)$$

$$\left\langle \widetilde{\Upsilon}_{\mu}(\overrightarrow{k},\omega)\widetilde{\Upsilon}_{\nu}(\overrightarrow{k}',\omega') \right\rangle = \frac{2k_B}{\gamma_1} \delta^0_{\mu\nu}{}^{\perp}\widetilde{T}\left(\overrightarrow{k} - \overrightarrow{k}'\right) \delta\left(\omega - \omega'\right), \quad (23)$$

$$\left\langle \widetilde{g}_i(\overrightarrow{k},\omega)\widetilde{g}_j(\overrightarrow{k}',\omega') \right\rangle = 2k_B \kappa^0_{ij} \widetilde{T}^*\left(\overrightarrow{k} - \overrightarrow{k}'\right) \delta\left(\omega - \omega'\right), \quad (24)$$

where

$$\begin{aligned}
\widetilde{T}\left(\overrightarrow{k} - \overrightarrow{k}'\right) &= T_0 \delta\left(\overrightarrow{k} - \overrightarrow{k}'\right) \\
&\quad - \frac{\alpha}{2qi}\left[\delta\left(\overrightarrow{k} - \overrightarrow{k}' - \overrightarrow{q}\right) - \delta\left(\overrightarrow{k} - \overrightarrow{k}' + \overrightarrow{q}\right)\right]
\end{aligned} \quad (25)$$

and

$$\begin{aligned}
\widetilde{T}^*\left(\overrightarrow{k} - \overrightarrow{k}'\right) &= T_0^2 \delta\left(\overrightarrow{k} - \overrightarrow{k}'\right) \\
&\quad - \frac{\alpha T_0}{qi}\left[\delta\left(\overrightarrow{k} - \overrightarrow{k}' - \overrightarrow{q}\right) - \delta\left(\overrightarrow{k} - \overrightarrow{k}' + \overrightarrow{q}\right)\right] \\
&\quad - \frac{\alpha^2}{4q^2}[\delta\left(\overrightarrow{k} - \overrightarrow{k}' - 2\overrightarrow{q}\right) - \delta\left(\overrightarrow{k} - \overrightarrow{k}'\right) \\
&\quad + \delta\left(\overrightarrow{k} - \overrightarrow{k}' + 2\overrightarrow{q}\right)]
\end{aligned} \quad (26)$$

It should be pointed out that in deriving these expressions it was convenient to replace z by $\frac{1}{q}\sin(qz)$ and then take the limit $q \to 0$ after performing the transform.

3. LIGHT SCATTERING SPECTRUM

For a plane wave electric field incident on a nonmagnetic, nonconducting and nonabsorbent medium with an average electric permittivity, ϵ_0, the scattered electric field, E_s, at a large distance, R, is proportional to the Fourier transform of its dielectric tensor fluctuations $\delta\epsilon^{(i,f)}(\overrightarrow{k},t)$ [22],

$$E_s(\overrightarrow{R},t) = \frac{\overrightarrow{k}_f^2 E_0 e^{i(\overrightarrow{k}_f \cdot \overrightarrow{R} - \omega_i t)}}{4\pi\epsilon_0 R}\delta\epsilon^{(i,f)}(\overrightarrow{k},t). \quad (27)$$

Here ω_i is the angular frequency of the incident field and $\overrightarrow{k} \equiv \overrightarrow{k}_i - \overrightarrow{k}_f$ is the difference between the wave vectors of the incident wave and the wave that reaches the detector. The scattering angle, ϕ, is the angle between $\overrightarrow{k}_i$ and $\overrightarrow{k}_f$ and is related to $k =\mid \overrightarrow{k}\mid$ by $k = (4\pi/\lambda_i)\sin(\phi/2)$, where λ_i is the wavelength of the incident field. The shifts in frequency will be denoted by $\omega \equiv \omega_f - \omega_i$, where ω_f is the frequency of the field at the detector. $E_0 = \left|\overrightarrow{E}_0\right|$ is the magnitude of the incident field and $\delta\epsilon^{(i,f)}(\overrightarrow{k},t) \equiv \widehat{i}\cdot \delta\overleftrightarrow{\epsilon}(\overrightarrow{k},t)\cdot \widehat{f}$, where $\widehat{i}$ and $\widehat{f}$ are the initial and final directions of the polarizations of the incident and scattered light.

The light scattering spectrum, $\chi(\overrightarrow{k},\omega)$, of a nematic liquid crystal is proportional to the dielectric tensor fluctuations autocorrelation function , [23], [22], namely,

$$\chi(\overrightarrow{k},\omega) = \frac{\overrightarrow{k}_f^4 |E_0|^2}{32\pi^3\epsilon_0^2 R^2}\int_{-\infty}^{\infty}e^{-i\omega t}\left\langle \delta\epsilon^{(i,f)}(\overrightarrow{k},t)\delta\epsilon^{(i,f)*}(\overrightarrow{k},0)\right\rangle dt, \quad (28)$$

where the asterisk denotes complex conjugate. Furthermore, since a nematic liquid crystals is an uniaxial media, its dielectric tensor is of the form

$$\epsilon_{\alpha\beta} = \epsilon_\perp \delta_{\alpha\beta} + \epsilon_a n_\alpha n_\beta, \quad (29)$$

where the dielectric anisotropy $\epsilon_a \equiv \epsilon_\parallel - \epsilon_\perp$, is the difference between the dielectric constants along the director and a direction perpendicular to it. Thus, for the present model one finds that

$$\left\langle \delta\epsilon^{(i,f)}(\overrightarrow{k},t)\delta\epsilon^{(i,f)*}(\overrightarrow{k},0)\right\rangle = \left\langle \delta n_y(k_y,k_z,t)\delta n_y^*(k_y,k_z,0)\right\rangle$$

and the spectrum is specified by only one correlation function

$$\begin{aligned}\chi(k_y,k_z,\omega) &= \frac{\overrightarrow{k}_f^4 |E_0|^2}{32\pi^3\epsilon_0^2 R^2}\int_{-\infty}^{\infty}e^{-i\omega t}\left\langle \delta n_y(k_y,k_z,t)\delta n_y^*(k_y,k_z,0)\right\rangle dt\\ &\equiv \frac{\overrightarrow{k}_f^4 |E_0|^2}{32\pi^3\epsilon_0^2 R^2}C(k_y,k_z,\omega).\end{aligned} \quad (30)$$

4.   CORRELATION FUNCTION

To calculate the relevant equilibrium correlation function $C(k_y,k_z,\omega)$ in (5), we first solve (16) for $\delta n_y(k_y,k_z,\omega)$ with the result

$$\delta n_y(k_y,k_z,\omega) = \frac{1}{U}[M_0 g_y + N_0 g_z + P_0\Sigma_1 + Q_0\Sigma_2 + R_0\Sigma_3 + S_0\Sigma_4 + W_0\Upsilon_y]. \quad (31)$$

The coefficients M_0, N_0, P_0, Q_0, R_0, S_0, W_0 and U, $U \equiv \det \mid M \mid$, are quite complicated functions of (k_y,k_z,ω) and their explicit forms are given in Appendix A. Using (31) and making use of the required fluctuation–dissipation relations given by (22)-(24), we may calculate $C(k_y, k_z, \omega)$. Note that when the temperature profile (2) is inserted into these fluctuation–dissipation relations, each relation has an equilibrium and a nonequilibrium contribution and

$$C(k_y, k_z, \omega) = C(k_y, k_z, \omega)^{eq} + C(k_y, k_z, \omega)^{neq}. \tag{32}$$

More explicitly,

$$C(k_y, k_z, \omega)^{eq} = \frac{2k_B T_0}{UU^*} \left\{ \left(\kappa_\perp M_0 M_0^* + \kappa_\parallel N_0 N_0^* \right) T_0 + (\nu_2 + \nu_4) P_0 P_0^* \right.$$
$$+ \eta_5 \left(S_0 P_0^* + P_0 S_0^* \right) + \eta_3 \left(Q_0 Q_0^* + Q_0 R_0^* + R_0 Q_0^* + R_0 R_0^* \right)$$
$$\left. + (2\eta_1 + \eta_2 - \eta_4 + 2\eta_5) S_0 S_0^* + \frac{1}{\gamma_1} W_0 W_0^* \right\}, \tag{33}$$

while the nonequilibrium contribution is

$$C(k_y, k_z, \omega)^{neq} = \frac{i\alpha}{2q} \frac{2k_B}{UU^*} [2T_0 \left(\kappa_\perp M_0 M_0^* + \kappa_\parallel N_0 N_0^* \right) + (\nu_2 + \nu_4) P_0 P_0^*$$
$$+ \eta_5 \left(S_0 P_0^* + P_0 S_0^* \right) + \eta_3 \left(Q_0 Q_0^* + Q_0 R_0^* + R_0 Q_0^* + R_0 R_0^* \right)$$
$$+ (2\eta_1 + \eta_2 - \eta_4 + 2\eta_5) S_0 S_0^*$$
$$+ \frac{1}{\gamma_1} W_0 W_0^*] \left[\delta \left(k_z - k_z' - q \right) - \delta \left(k_z - k_z' + q \right) \right]. \tag{34}$$

The complicated explicit forms of the coefficients $M_0 M_0^*$, $N_0 N_0^*$, $P_0 P_0^*$, $Q_0 Q_0^*$, $R_0 R_0^*$, $R_0 Q_0^*$, $Q_0 R_0^*$, $S_0 P_0^*$, $P_0 S_0^*$, UU^* as functions of (k_y, k_z, ω), are obtained from the expressions in Appendix; they are given explicitly in Ref. [21]. To linear order in α, the right hand side of (34) contains all the information about the nonequilibrium effects produced by the external thermal gradient in the dynamic structure factor.

5.   RESULTS

It is evident from Eqs. (33) and (34) that their explicit k dependence is quite complicated. However, in spite of their complexity the following overall k dependence may be established: $C^{eq} \sim k^{-6}$ and $C^{neq} \sim k^{-7}$, a result that reflects the fact that δn_y is not a conserved quantity. For the transverse variables the corresponding dependence were $C^{eq} \sim k^{-4}$ and $C^{neq} \sim k^{-5}$, [13]. Also, in Ref. [13] the Brillouin peaks of the structure factor of a simple fluid under a temperature gradient, which only involve fluctuations of conserved variables, the k dependence of the equilibrium and nonequilibrium contributions to the spectrum goes as k^{-2} and k^{-4}, respectively, [1], [2].

If we normalize $\chi(k_y, k_z, \omega)$ with the static correlation $\chi^{eq}(\vec{k}, 0)$, the normalized spectrum $S(k_y, k_z, \omega) = S^{eq}(k_y, k_z, \omega) + S^{neq}(k_y, k_z, \omega)$ may be calculated. To this end the dispersion equation, det $M_{ij} = U(k_y, k_z, \omega)$ $= 0$, should be solved. This is a quintic equation whose explicit form is also given in Appendix A and its solution yields the hydrodynamic modes of the system. By following the standard procedure of assuming that ω is an analytic function of $\vec{k}$ and by writing a series of perturbation equations for the zeroth, first, ..., n^{th} order equations in the small quantities $D_T k^2$, $D_\eta k^2$, $D_\gamma k^2$, $D_\sigma k^2$, $D_\xi k^2$, $D_\nu k^2$, which represent the inverse characteristic times associated with the relaxation processes occurring in the nematic such as thermal diffusivity (D_T), shear viscosity (D_η), orientational relaxation (D_γ), etc. Here $D_V \equiv D_\xi + D_\nu$, $D_\eta = k^{-2}(D_\nu k_y^2 - 2D_\mu k_y k_z + D_\xi k_z^2)$ and $D_\sigma \equiv \frac{\gamma_1}{\rho_0 k^4} \left(\lambda_0 k_y^2 - \lambda_1 k_z^2\right)^2$. In this way we find the following roots of the dispersion relation corresponding to the longitudinal modes [21], [24]. The thermal diffusivity mode

$$\omega_T = -iD_T k^2, \tag{35}$$

the shear mode

$$\omega_v = -iD_\eta k^2, \tag{36}$$

the diffusion mode induced by the relaxation of the director

$$\omega_n = -iD_\gamma k^2 (1 + \frac{D_\sigma}{D_\eta}) \tag{37}$$

and the two sound modes

$$\omega_{s+,-} = \pm c_s k - i\Gamma k^2, \tag{38}$$

where

$$\Gamma \equiv \frac{1}{2} \left[(\gamma - 1) D_T + D_V + D_\eta\right]. \tag{39}$$

¿From these expressions it follows that the first three diffusion modes show a substantial anisotropy and that the sound modes, $\omega_{s+,-}$, show an anisotropic sound attenuation, but the sound velocity is isotropic.

In order to plot the light scattering spectrum we consider the thermotropic $MBBA$, which is nematic in the temperature range $20°C - 47°C$. The required material parameters are given in table 1.

Using (33) we find that $S^{eq}(k_y, k_z, \omega)$ as a function of ω is given by the curve in Figure 2. The nonequilibrium contribution $S^{neq}(k_y, k_z, \omega)$ was evaluated in the following approximate way due to the complexity of (34). In this equation we only considered the first term on the right hand side which contains the contribution due to the fluctuations in the

λ	1.03
ρ_0	$1.029\ g\ cm^{-3}$
ν_2	$41.6 \times 10^{-2}\ g\ cm^{-1}s^{-1}$
ν_3	$23.8 \times 10^{-2}\ g\ cm^{-1}\ s^{-1}$
γ_1	$76.3 \times 10^{-2} g\ cm^{-1}\ s^{-1}$
K_2	$2.2 \times 10^{-7}\ g\ cm\ s^{-2}$
K_3	$7.45 \times 10^{-7}\ g\ cm\ s^{-2}$

Table 1.   Material parameters for $MBBA$ [17]

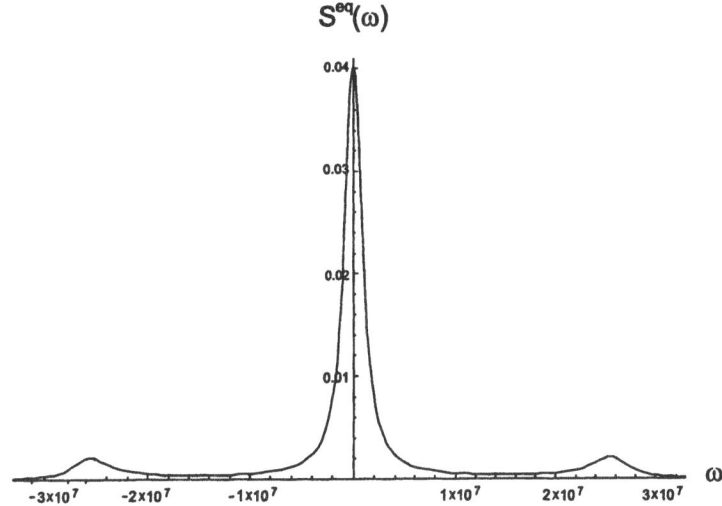

Figure 2.   The equilibrium structure factor $S^{eq}(\omega)$ plotted as a function of ω (arbitrary units) for $MBBA$. Here the dimensionless temperature gradient is $\overline{\alpha} \equiv 0.5$, $T_0 = 293\ K$, $d = 10^{-2}\ cm$ and $\phi \sim 1°$.

heat flux, g_y, and this leads to the plot in Figure 3. The effect of both contributions are plotted in Figure 4. It shows that both, the central and the Brillouin peaks are affected by the thermal gradient. Since this later distortion is hardly appreciated from Figure 4, in Figures 5 and 6 these effects on the sound peaks are amplified. The left hand side peak diminishes while the right hand side one increases by the same amount due to the thermal gradient. This effect is similar to the one observed for a simple fluid in Ref. [8]. But the central part of the spectrum produced by the modes ω_T, ω_v and ω_n, shows a complicated distortion with respect to the equilibrium contribution. It should be emphasized, though, that we have only considered the first term in (34), but this shows already the tendency induced by the thermal steady state on the dynamic structure factor.

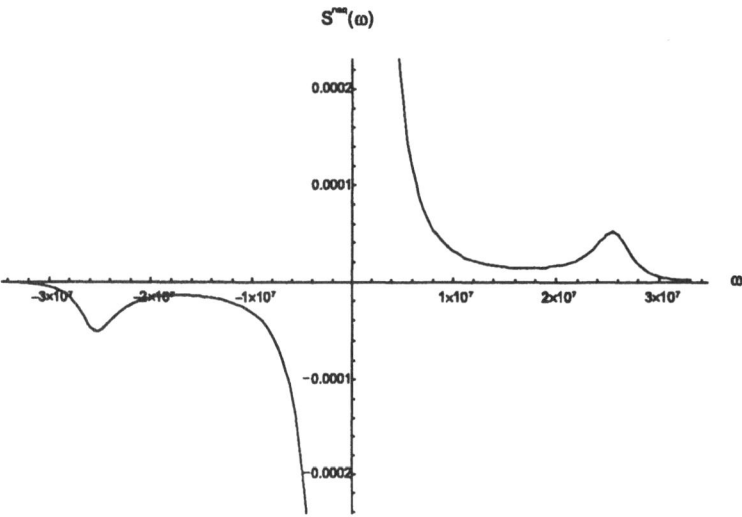

Figure 3.   The same as in Figure 2 for $S^{neq}(\omega)$.

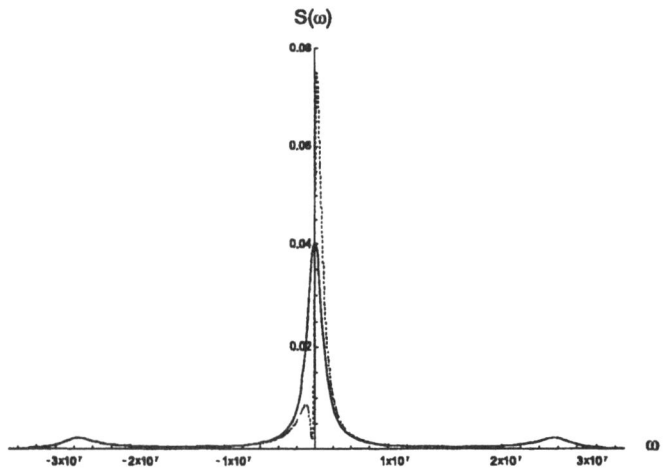

Figure 4.   The total structure factor $S(\omega)$ plotted as a function of ω for $MBBA$ and for the same parameter values as in Figure 2.

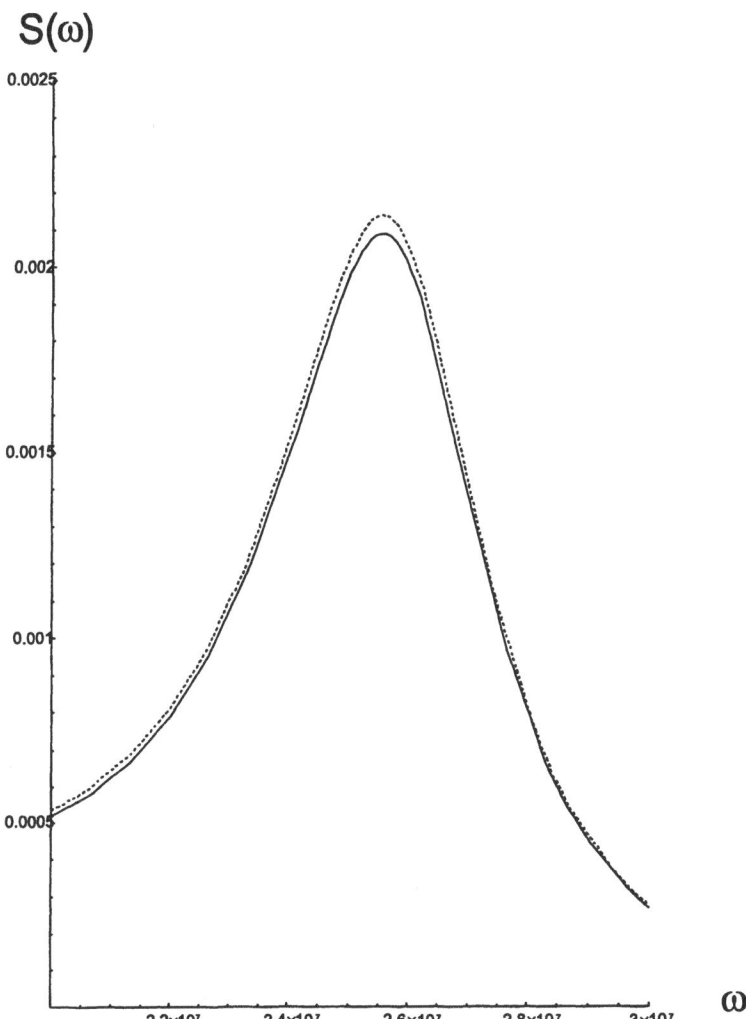

Figure 5.   Amplification of the right hand side Brillouin peak of the spectrum shown in Figure 4.

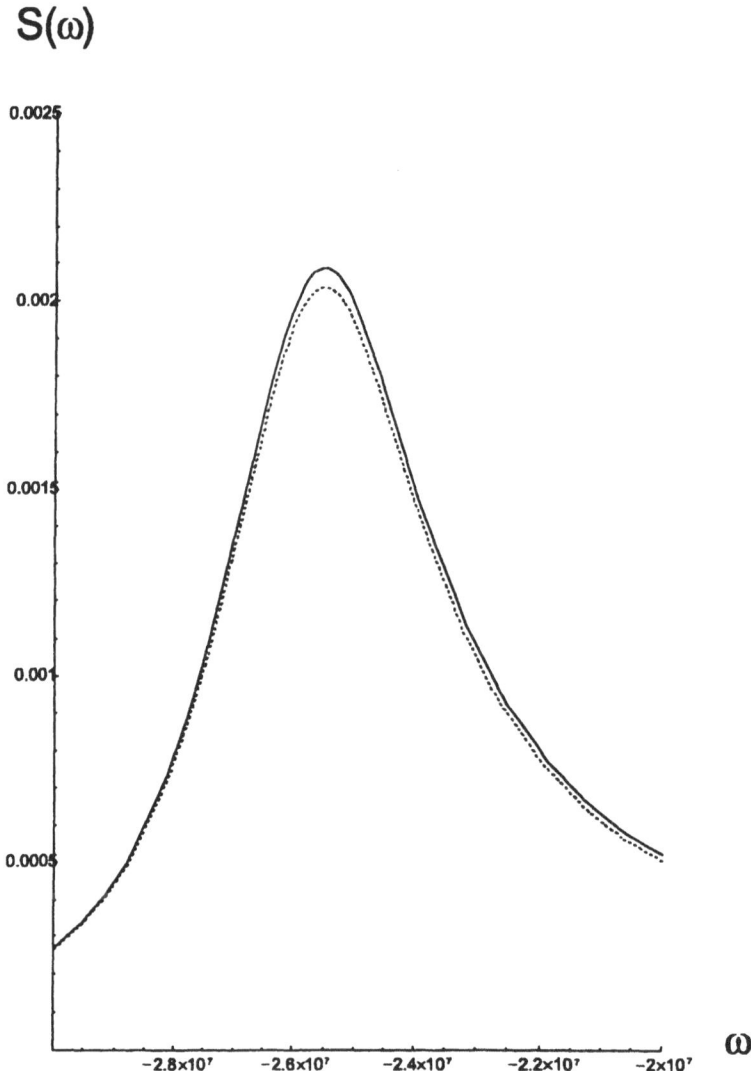

Figure 6.   The same as in Figure 5 for the left Brillouin peak.

6.   CONCLUSIONS

Summarizing, we have investigated theoretically the influence of the effects produced by an external temperature gradient on the light scattering spectrum of a thermotropic nematic due to the longitudinal hydrodynamic modes. This problem is a partial analog for liquid crystals of the situation considered in references [1], [2], [8] for simple fluids.

For the model considered, the nonequilibrium correction is an odd function of ω that introduces an asymmetry in the shape of the structure factor; it increases the intensity of the central peak and shifts its maximum towards positive values of ω. Although it was not shown analytically, the size of the shift depends on the magnitude of the gradient, as indicated in Figure 4. On the other hand, the presence of the thermal gradient also affects the shape of the Brillouin spectrum, since one of the peaks shrinks in the same amount as the other increases.

The differences in the found k dependence, $\chi^{eq} \sim k^{-6}$ and $\chi^{neq} \sim k^{-7}$, as compared to the dependence of the transverse variables (δn_x), $C^{eq} \sim k^{-4}$ and $C^{neq} \sim k^{-5}$, or for the Brillouin peaks in the nonequilibrium state generated by a temperature gradient, $\chi^{eq} \sim k^{-2}$ and $\chi^{neq} \sim k^{-4}$, are related to the nature of the variables involved. For a simple fluid all the variables are conserved, whereas for the liquid crystal δn_x or δn_y are not conserved variables.

It should be stressed that our estimation of the nonequilibrium effect is only a crude approximation since we have only taken into account in (34) only part of the contribution due to the fluctuations in the heat flux, g_y, and we ignored all the fluctuations in the stress tensor and the current associated with the relaxation of the director. A more complete analysis should include all these contributions and is presently under way [24]; however, our results show the tendency of the changes produced on the spectrum by the external thermal gradient.

To our knowledge, there are no experimental data available in the literature to check our results. Whether the effects due to convection may be ignored remains to be assessed. However, the model has been constructed so that it corresponds to a homodyne experimental arrangement and appropriate to detect the so called mode 1 of the spectrum [23].

Acknowledgments

We acknowledge partial financial support from grants $DGAPA - UNAM\ IN$ 101999 and FENOMEC–CONACYT 400316–5–G25427E, Mexico.

Appendix

In this Appendix we list the expressions of the coefficients in Eq. (33) as functions of (k_y, k_z, ω) as defined in Ref. [21].

$$
\begin{aligned}
M_0(k_y, k_z, \omega) = \ & -\gamma\theta\left(\lambda_1 + \lambda_0\right) k_y^2 k_z \omega^2 + i\gamma\theta[D_\mu\left(\lambda_0 k_y^2 + \lambda_1 k_z^2\right) k_y \\
& -\left(\lambda_0 D_\xi + \lambda_1 D_\nu\right) k_y^2 k_z] k^2 \omega,
\end{aligned} \tag{A.1}
$$

$$
\begin{aligned}
N_0(k_y, k_z, \omega) = \ & -\gamma\theta\left(\lambda_0 + \lambda_1\right) k_y k_z^2 \omega^2 \\
& +i\gamma\theta\left[D_\mu\left(\lambda_0 k_y^2 + \lambda_1 k_z^2\right) k_z - \left(\lambda_0 D_\xi + \lambda_1 D_\nu\right) k_y k_z^2\right] k^2 \omega
\end{aligned} \tag{A.2}
$$

$$
\begin{aligned}
P_0(k_y, k_z, \omega) = \ & -\gamma\lambda_1 k_z k_y \omega^3 + i\gamma\left(\lambda_0 D_\mu k_y^2 - \lambda_1 D_\nu k_z k_y - \lambda_1\gamma D_T k_z k_y\right) k^2 \omega^2 \\
& +\gamma\left[c_s^2\left(\lambda_1 k_z^2 - \lambda_0 k_y^2\right) k_y k_z + \gamma D_T\left(\lambda_1 D_\nu k_z k_y - \lambda_0 D_\mu k_y^2\right) k^4\right]\omega \\
& +i c_s^2 \gamma D_T\left(\lambda_1 k_z^2 - \lambda_0 k_y^2\right) k_y k_z k^2,
\end{aligned} \tag{A.3}
$$

$$
\begin{aligned}
Q_0(k_y, k_z, \omega) = \ & -\gamma\lambda_1 k_z^2 \omega^3 + i\gamma\left(\lambda_0 D_\mu k_y k_z - \lambda_1 D_\nu k_z^2 - \lambda_1\gamma D_T k_z^2\right) k^2 \omega^2 \\
& +\gamma\left[c_s^2\left(\lambda_1 k_z^2 - \lambda_0 k_y^2\right) k_z^2 + \gamma D_T\left(\lambda_1 D_\nu k_z^2 - \lambda_0 D_\mu k_y k_z\right) k^4\right]\omega \\
& +i c_s^2 \gamma D_T\left(\lambda_1 k_z^2 - \lambda_0 k_y^2\right) k_z^2 k^2,
\end{aligned} \tag{A.4}
$$

$$
\begin{aligned}
R_0(k_y, k_z, \omega) = \ & -\gamma\lambda_0 k_y^2 \omega^3 + i\gamma\left(\lambda_1 D_\mu k_z k_y - \lambda_0 D_\xi k_y^2 - \lambda_0\gamma D_T k_y^2\right) k^2 \omega^2 \\
& +\gamma\left[c_s^2\left(\lambda_0 k_y^2 - \lambda_1 k_z^2\right) k_y^2 + \gamma D_T\left(\lambda_0 D_\xi k_y^2 - \lambda_1 D_\mu k_z k_y\right) k^4\right]\omega \\
& +i c_s^2 \gamma D_T\left(\lambda_0 k_y^2 - \lambda_1 k_z^2\right) k_y^2 k^2,
\end{aligned} \tag{A.5}
$$

$$
\begin{aligned}
S_0(k, \omega) = \ & -\gamma\lambda_0 k_y k_z \omega^3 + i\gamma\left(\lambda_1 D_\mu k_z^2 - \lambda_0 D_\xi k_y k_z - \lambda_0\gamma D_T k_y k_z\right) k^2 \omega^2 \\
& +\gamma\left[c_s^2\left(\lambda_0 k_y^2 - \lambda_1 k_z^2\right) k_y k_z + \gamma D_T\left(\lambda_0 D_\xi k_y k_z - \lambda_1 D_\mu k_z^2\right) k^4\right]\omega \\
& +i c_s^2 \gamma D_T\left(\lambda_0 k_y^2 - \lambda_1 k_z^2\right) k_y k_z k^2
\end{aligned} \tag{A.6}
$$

$$
\begin{aligned}
W_0(k_y, k_z, \omega) = \ & \gamma\rho_0[-i\omega^4 + \left(D_V + \gamma D_T\right) k^2 \omega^3 \\
& +i\left(k^2 c_s^2 + D_\Theta k^4 + \gamma D_T D_V k^4\right)\omega^2 \\
& -\left(c_s^2 D_T k^4 + \gamma D_T D_\Theta k^6 + c_s^2 D_\eta k^4\right)\omega \\
& -i c_s^2 D_T D_\eta k^6].
\end{aligned} \tag{A.7}
$$

The dispersion equation is given by

$$U(k,\omega) \equiv \gamma\rho_0\{\omega^5 + i\left(\gamma D_T + D_\nu + D_\xi + D_\gamma\right)k^2\omega^4$$

$$-[\gamma D_T\left(D_\xi + D_\nu\right)k^4 + \gamma D_T D_\gamma k^4 + D_\gamma\frac{\gamma_1}{\rho_0}\left(\lambda_0^2 k_y^2 + \lambda_1^2 k_z^2\right)k^2 + \left(D_\nu D_\xi - D_\mu^2\right)k^4$$

$$+D_\gamma\left(D_\xi + D_\nu\right)k^4 + k^2 c_s^2]\omega^3 - i[\gamma D_T D_\gamma\frac{\gamma_1}{\rho_0}\left(\lambda_1^2 k_z^2 + \lambda_0^2 k_y^2\right)k^4$$

$$+\gamma D_T\left(D_\nu D_\xi - D_\mu^2\right)k^6 + \gamma D_T D_\gamma\left(D_\nu + D_\xi\right)k^6 + D_\gamma\left(D_\nu D_\xi - D_\mu^2\right)k^6$$

$$+D_\gamma\frac{\gamma_1}{\rho_0}\left(\lambda_1^2 D_\nu k_z^2 - 2\lambda_0\lambda_1 D_\mu k_z k_y + \lambda_0^2 D_\xi k_y^2\right)k^4 + c_s^2\left(D_\nu k_y^2 + 2D_\mu k_y k_z - D_\xi k_z^2\right)k^2$$

$$+c_s^2\left(D_\gamma + D_T\right)k^4]\omega^2 + [\gamma D_T D_\gamma\left(D_\nu D_\xi - D_\mu^2\right)k^8$$

$$+\gamma D_T D_\gamma\frac{\gamma_1}{\rho_0}\left(\lambda_1^2 D_\nu k_z^2 - 2\lambda_1\lambda_0 D_\mu k_z k_y + \lambda_0^2 D_\xi k_y^2\right)k^6$$

$$+c_s^2 D_T D_\gamma k^6 + c_s^2\left(D_\gamma + D_T\right)\left(D_\xi k_z^2 - 2D_\mu k_y k_z + D_\nu k_y^2\right)k^4$$

$$+c_s^2 D_\gamma\frac{\gamma_1}{\rho_0}\left(\lambda_0 k_y^2 - \lambda_1 k_z^2\right)^2 k^2]\omega$$

$$+ic_s^2[D_T D_\gamma\left(D_\nu k_y^2 - 2D_\mu k_y k_z k^6 + D_\xi k_z^2\right)k^6$$

$$+D_T D_\gamma\frac{\gamma_1}{\rho_0}\left(\lambda_0 k_y^2 - \lambda_1 k_z^2\right)^2 k^4]\} = 0. \tag{A.8}$$

References

[1] D. Ronis, I. Procaccia and J. Machta, *Phys. Rev.* **A22** (1980) 714.

[2] A. M. S. Tremblay, E. D. Siggia and M. R. Arai, *Phys. Lett.* **A76** (1980) 57.

[3] J. Machta, I. Oppenheim and I. Procaccia, *Phys. Rev.* **A22** (1980) 2809.

[4] L. S. García-Colín and R. M. Velasco, *Phys. Rev.* **A12** (1975) 646.

[5] R. M. Velasco and L. S. García–Colín, *J. Phys.* **A24** (1991) 1007.

[6] J. M. Ortíz de Zárate and J. V. Sengers, *"Boundary effects on the nonequilibrium structure factor of fluids below the Rayleigh–Bénard instability"*, preprint (2002).

[7] J. Bonet-Avalos, J. M. Rubí, R. F. Rodríguez and A. Pérez–Madrid, *Phys. Rev.* **A41** (1990) 1923.

[8] D. Beysens, Y. Garrabos and G. Zalczer, *Phys. Rev. Lett.* **45** (1980) 403.

[9] D. Beysens, in: *Scattering Techniques Applied to Supramolecular and Non-Equilibrium Systems*, S. H. Shang, B. Chu and R.Nossal, eds. (Plenum Press, New York, 1981) pp. 537.

[10] H. Kiefte, M. J. Clouter and R. Penney, *Phys. Rev.* **B30** (1984) 4017.

[11] R. N. Suave and A. R. B. de Castro, *Phys. Rev.* **B53** (1996) 5330.

[12] H. F. Gleeson, in: *Handbook of Liquid Crystals*, D. Demus, J. Goodby, G. W. Gray, H. W. Spiess and V. Vill, eds. (Wiley–Vch, Weiheim, 1998).

[13] H. R. Brand and H. Pleiner, *Phys. Rev.* **A35** (1987) 3122.

[14] H. Pleiner and H. R. Brand, *J. Physique* **44** (1983) L–23.

[15] R. F. Rodríguez and J. F. Camacho, *Rev. Mex. Fís.* **48** (2002) (in print).

[16] D. Foster, *Hydrodynamic Fluctuations, Broken Symmetry and Correlation Functions* (Benjamin, Reading, 1975).

[17] P. G. de Gennes, *The Physics of Liquid Crystals* (Oxford University Press, Oxford, 1974).

[18] H. Pleiner and H. R. Brand, in: *Pattern Formation in Liquid Crystals*, A. Buka and L. Kramer, eds. (Springer, New York, 1996).

[19] L. D. Landau and E. Lifshitz, *Fluid Dynamics* (Pergamon, New York, 1959).

[20] R. F. Fox, *Phys. Rep.* **48** (1978) 179.

[21] J. F. Camacho, *Ph.D. Dissertation*, Universidad Nacional Autónoma de México (2002) (in Spanish).

[22] B. J. Berne and R. Pecora, *Dynamic Light Scattering* (Krieger, Malabar, 1990).

[23] *Handbook of Liquid Crystals*, D. Demus, J. Goodby, G. W. Gray, H. W. Spiess and V. Vill, editors (Wiley–Vch, Weiheim, 1998).

[24] J. F. Camacho and R. F. Rodríguez (2002) (unpublished).

RELAXATION IN THE KINETIC ISING MODEL ON THE PERIODIC INHOMOGENEOUS CHAIN

L. L. Goncalves
Departamento de Fisica, Universidade Federal do Ceará
Campus do Pici, C. P. 6030, 60451-970, Fortaleza, Ceará, Brazil
lindberg@fisica.ufc.br

M. López de Haro
Centro de Investigación en Energía, U.N.A.M.
Apdo. Postal 34, Temixco, Mor. 62580, México
malopez@servidor.unam.mx

J. Tagüeña–Martinez
Centro de Investigación en Energía, U.N.A.M.
Apdo. Postal 34, Temixco, Mor. 62580, México
jtag@servidor.unam.mx

A. P. Vieira
Instituto de Fisica, Departamento de Fisica General
Universidade de São Paulo
C.P. 66318, 05315-970, São Paulo, SP, Brazil
apvieira@if.usp.br

Abstract   The kinetic Ising model on an inhomogeneous periodic chain, which is composed of N segments with n different bonds, is considered. The model is studied within the Glauber dynamics and the restricted dynamics introduced by the authors for the description of relaxation phenomena in linear polymer chains, and it corresponds to an extension of the alternating bond model and the n–isotopic model. For both dynamics, the solution of the model may be obtained exactly for arbitrary

Developments in Mathematical and Experimental Physics,
Volume B: Statistical Physics and Beyond
Edited by Macias *et al.*, Kluwer Academic/Plenum Publishers, 2003

n. The dynamical critical exponent *z* and the response function are determined and the universality of both dynamics is discussed.

Keywords: Kinetic Ising model, periodic inhomogeneous chain, Glauber dynamics, restricted dynamics, dynamic critical exponent.

1.   INTRODUCTION

The work reported in this paper continues a line of research started some twelve years ago when we became interested in studying some aspects related to glass-forming systems. As it is well known, the properties of materials do not depend on their chemical constitution only, but also on the arrangement and the dynamics of the molecules. Cooling a glass–forming substance preserves its liquid properties until, ultimately, the structure becomes frozen in a particular configuration. This configuration represents only one of the many subminima of a very rich energetic landscape. The system is now incapable of exploring its energy phase space within finite times. This frozen disordered structure is not in a global thermodynamic equilibrium and memory effects are fundamental. From a geometrical point of view, both liquid and glassy states are disordered. However, liquids have low viscosity and the disordered configuration changes very quickly. On the other hand, there are many glassy systems that combine an irregular structure with extremely high values of viscosity and maintain their disordered liquid like structure upon cooling forming a new state. Irrespective of their chemical nature, most glasses display similar features absent in their counterparts in crystalline materials and are systems in which the time scale of positional changes is much longer than the experimental times. Therefore, disorder appears to be a dominant factor leading to these characteristics.

Phenomenologically, there is a gradual transition from the liquid to the solid state, which makes it difficult to determine a clear–cut transition temperature T_g that discriminates between the liquid and the glassy states. In fact, T_g, which depends on the cooling or heating rates, represents the temperature where the viscous flow stops and solidification must be seen as a dynamic property linked to the time scale and to the specific relaxation under study.

A vast number of methods exists to disturb a system in order to monitor its relaxation towards equilibrium. Some of them are nuclear magnetic resonance techniques, mechanical, dielectric, enthalpy and volume relaxation. Relaxations that go hand in hand with solidification are usually referred to as α–processes. Secondary relaxations, referred to as β –processes, can occur over a wide range of temperatures and high frequencies. The dielectric spectroscopy plays an important role in the investigation of the dynamics of the molecules in glasses due to the exceptionally wide frequency window accessible with this method. The broad–band dielectric spectra reveal the large variety of processes governing the dynamic response mentioned above. Unfortunately, relaxation experiments cannot differentiate between the factors and are not enough to explore exhaustively the microscopic origin of an observed decayed process. Furthermore, in glassy systems, not only the temporal aspect of the relaxation counts but also the temperature dependence of the underlying rates seems to obey general equations. As universal features of disordered systems one finds relaxation patterns which depend on time and temperature in a complex way. Exponential response functions and Arrhenius–like temperature behavior are no longer the rule. What is more surprising is not the deviations from simple rules in amorphous materials, but often the time and temperature dependence of relaxations turns out to be similar, even for glasses which differ strongly in their chemical and physical constitution [1].

A few years ago, while trying to describe experimental work on dielectric relaxation in glass–forming liquids, Nagel and coworkers [2] proposed a (universal) scaling hypothesis to replace the usual normalized Debye scaling. Such a hypothesis has had an undeniably great phenomenological success but remains yet to be fully understood. The question of whether most properties of glasses and the glass transition phenomenon behave with universal features or not has been a source of disputes for many years. The use of the technical term 'universality' for glasses is to be understood in a different way compared to the case of the usual phase transitions in statistical mechanics. In this latter case, the physical situation is simple: although of course the location of the critical point itself depends on the actual system, most of the scaling properties of critical quantities, such as the order parameter and the correlation length, and all the critical exponents depend on the dimension of the order parameter and on the dimensionality rather than on the details of the materials or the precise form of the interaction energy. In glass transition phenomena the situation is clearly not at all simple and the term universality should therefore not be used, at least not in a similar context as in field theory.

Once the term 'universality' is understood in the way just discussed, in order to gain insight into the intricacies of glass–forming materials but avoiding the difficulties as much as possible, it is natural to investigate simple models that allow a glass-like transition. A further motivation in this respect is the fact that recently we pointed out the possible connection between the Nagel scaling hypothesis and microscopic models [3, 6]. The models we considered were kinetic Ising models in linear chains either using the Glauber dynamics [7], or a restricted dynamics [8], which we have introduced for the description of relaxation phenomena in linear polymer chains. Of course these models are highly oversimplified to be compared in detail with realistic systems but, apart from allowing detailed analytical computations, they attempt to incorporate the features that seem to be essential to account for the various experimental observations. In this regard we wholly subscribe the following statement taken from a paper by Angles d'Auriac and Rammal [9]: "Kinetic Ising models correspond to the simplest dynamical pictures of real systems and are believed to provide a sensible description of the dynamics of a large number of physical systems". It is the major aim of this paper to make use of the general method we have followed to study relaxation properties of kinetic Ising models on linear chains in order to illustrate its simplicity and versatility as well as to provide some new results on the dynamical critical exponent and the response function of the periodic inhomogeneous kinetic Ising model. This model corresponds to an extension of the alternating bond model [10] and in a way attempts to account for some kind of disorder. It is worth noting that the study of the critical dynamics of the inhomogeneous kinetic Ising model, in linear chains, has attracted considerable attention in recent years since this is the first model where it has been shown rigorously that, within the Glauber dynamics [7], the universality of the dynamic scaling hypothesis breaks down [11]. The model has also been studied in the framework of the restricted dynamics [4], [5] where it has been shown that, at least for the alternating bond chain, universality is preserved.

2. THE KINETIC ISING MODEL ON THE PERIODIC INHOMOGENEOUS LINEAR CHAIN

Here we consider the Ising model on a linear chain with N identical segments, each one containing n sites occupied by isotopes characterized by n different spin relaxation times. Allowing also for different coupling strengths between the spins, the Hamiltonian for the lth segment is given by

$$H_l = -\sum_{j=1}^{n} J_j \sigma_{l,j} \sigma_{l,j+1}, \tag{1}$$

where $\sigma_{l,j}$ is a stochastic (time-dependent) spin variable taking the values ± 1, so that $\sigma_{l,j}^2 = 1$. The configuration of the segment is specified by the set of variables $\{\sigma^{l,n}\} \equiv \{\sigma_{l,1}, \sigma_{l,2}, \cdots, \sigma_{l,n}\}$ at time t, and it evolves in time due to interactions with a heat bath. It should be emphasized that we specify the Hamiltonian for each segment to have a precise definition of the states that intervene in allowed transitions, but it will play no further relevant role in the calculations that follow.

2.1. GLAUBER DYNAMICS

We assume for each segment the usual Glauber dynamics so that the transition probabilities are given by

$$w_{li}(\sigma_{l,i}) = \alpha_i \left(1 - \tanh \beta \left(J_{i-1}\sigma_{l,i-1}\sigma_{l,i} + J_i\sigma_{l,i}\sigma_{l,i+1}\right)\right), \qquad (2)$$

where $\beta = 1/k_B T$, k_B being the Boltzmann constant and T the absolute temperature, and α_i is the inverse of the relaxation time τ_i of spin i in the absence of spin interactions. This kind of dynamics ensures that detailed balance holds and hence an equilibrium state is finally attained in the model.

The time dependent probability $P\left(\left\{\sigma^{l,n}\right\}, t\right)$ for a given spin configuration within each segment satisfies the master equation

$$
\begin{aligned}
\frac{dP\left(\left\{\sigma^{l,n}\right\}, t\right)}{dt} &= -\sum_{i=1}^{n} w_{li}\left(\sigma_{l,i}\right) P\left(\left\{\sigma^{l,n}\right\}, t\right) \\
&\quad + \sum_{i=1}^{n} w_{li}\left(-\sigma_{l,i}\right) P\left(T_{li}\left\{\sigma^{l,n}\right\}, t\right),
\end{aligned}
\qquad (3)
$$

where $T_{li}\left\{\sigma^{l,n}\right\} \equiv \left\{\sigma_{l,1}, \sigma_{l,2}, \cdots, \sigma_{l,i-1}, -\sigma_{l,i}, \sigma_{l,i+1}, \cdots, \sigma_{l,n}\right\}$. The dynamical properties we are interested in, namely the dynamical critical exponent and the susceptibility, require the knowledge of some moments of the probability $P\left(\left\{\sigma^{l,n}\right\}, t\right)$. Hence, we introduce the following expectation values defined as:

$$q_{l,i}(t) = \langle \sigma_{l,i}(t) \rangle = \sum_{\left\{\sigma^{l,n}\right\}} \sigma_{l,i} P\left(\left\{\sigma^{l,n}\right\}, t\right). \qquad (4)$$

Multiplying the master equation by the appropriate quantities and performing the required summations we obtain the set of time evolution equations that will be used in our later development. These are given by

$$\frac{dq_{l,m}}{dt} = -\alpha_m \left[q_{l,m} - \langle \tanh \beta \left(J_{m-1}\sigma_{l,m-1} + J_m\sigma_{l,m+1}\right)\rangle\right]. \qquad (5)$$

Owing to the periodic structure of the chain, we must have $J_0 \equiv J_n$, $\sigma_{l,0} \equiv \sigma_{l-1,n}$ and $\sigma_{l,n+1} \equiv \sigma_{l+1,1}$. By expanding the hyperbolic tangent, the above equation can be rewritten as

$$\frac{dq_l}{dt} = -\alpha_m \left[q_{l,m} - \frac{\gamma_m^+}{2}q_{l,m-1} - \frac{\gamma_m^-}{2}q_{l,m+1}\right], \qquad (6)$$

where

$$\gamma_m^\pm = \tanh\left(\beta J_{m-1} + \beta J_m\right) \pm \tanh\left(\beta J_{m-1} - \beta J_m\right). \qquad (7)$$

After introducing the Fourier transforms

$$\hat{q}_{Q,m} = \frac{1}{\sqrt{N}} \sum_{l=1}^{N} e^{iQnl} q_{l,m}, \qquad (8)$$

where $Q = 2\pi\kappa/Nn$ ($\kappa = 0, 1, 2, ..., N$) and we have assumed a unit lattice constant, we obtain the differential equation

$$\frac{d\Psi_Q}{dt} = \mathbf{M}_Q \Psi_Q, \qquad (9)$$

where $\Psi_Q = (\hat{q}_{Q,1}, \hat{q}_{Q,2}, \cdots, \hat{q}_{Q,n})$ and $\mathbf{M}_Q$ is the matrix

$$\begin{pmatrix}
-\alpha_1 & \frac{\alpha_1 \gamma_1^-}{2} & 0 & \cdots & 0 & 0 & \frac{\alpha_1 \gamma_1^+}{2} e^{iQn} \\
\frac{\alpha_2 \gamma_2^+}{2} & -\alpha_2 & \frac{\alpha_2 \gamma_2^-}{2} & \cdots & 0 & 0 & 0 \\
0 & \frac{\alpha_3 \gamma_3^+}{2} & -\alpha_3 & \cdots & 0 & 0 & 0 \\
\vdots & \vdots & \vdots & \ddots & \vdots & \vdots & \vdots \\
0 & 0 & 0 & \cdots & -\alpha_{n-2} & \frac{\alpha_{n-2} \gamma_{n-1}^-}{2} & 0 \\
0 & 0 & 0 & \cdots & \frac{\alpha_{n-1} \gamma_{n-1}^+}{2} & -\alpha_{n-1} & \frac{\alpha_{n-1} \gamma_{n-1}^-}{2} \\
\frac{\alpha_n \gamma_n^-}{2} e^{-iQn} & 0 & 0 & \cdots & 0 & \frac{\alpha_n \gamma_n^+}{2} & -\alpha_n
\end{pmatrix}.$$

$$(10)$$

The solution to the differential equation is simply

$$\Psi_Q(t) = \exp(\mathbf{M}_Q t)\Psi_Q(0), \qquad (11)$$

and so the relaxation process of the Q-dependent magnetization is determined by the eigenvalues of $\mathbf{M}_Q$, which we denote by $\lambda_{Q,m}$ ($m = 1, 2, \cdots, n$).

The scaling hypothesis of Halperin and Hohenberg [12] relates the time scale τ_Q and the correlation length ξ through $\frac{1}{\tau_Q} \sim \xi^{-z} f(\xi Q)$ where f is the scaling function and z is the dynamical critical exponent. In the present model z is related to the critical mode $\lambda_{Q,1}$ (which goes to zero as $T \to 0, Q \to 0$) by the equation

$$\ln\left(-\frac{1}{\lambda_{Q,1}}\right) = c + \frac{2\tilde{J}z}{k_B T}, \qquad (12)$$

where c is an irrelevant constant, the correlation length ξ has been identified as $\xi \sim e^{\frac{\tilde{J}}{k_b T}}$ and $\tilde{J}$ is in general determined from the choice of coupling constants, leading to a non-universality of z.

The response function $S_Q(\omega)$, where ω is the frequency, is given by

$$S_Q(\omega) = \frac{1}{k_B T} \frac{1}{n} \sum_{m,m'=1}^{n} \left[\langle \hat{\sigma}_{-Q,m} \hat{\sigma}_{Q,m'} \rangle_{\mathrm{eq}} - i\omega \tilde{C}_Q(\omega) \right], \qquad (13)$$

where $\tilde{C}_Q(\omega)$ is the spatiotemporal Fourier transform of the two-spin correlation function,

$$\tilde{C}_Q(\omega) = \lim_{t'\to\infty} \frac{1}{2\pi n} \sum_{j,j'=1}^{n} \int_{-\infty}^{+\infty} dt \, \langle \hat{\sigma}_{-Q,j}(t') \hat{\sigma}_{Q,j'}(t'+t) \rangle e^{-i\omega t}, \quad (14)$$

which can be rewritten as

$$\tilde{C}_Q(\omega) = \sum_{j=1}^{n} \frac{g_{Q,j}}{i\omega - \lambda_{Q,j}}, \qquad (15)$$

where

$$g_{Q,j} = \sum_{m_1,m_2,m_3=1}^{n} \frac{\bar{a}_{Q,j,m_2} a_{Q,m_3,j}}{n} \langle \hat{\sigma}_{-Q,m_1} \hat{\sigma}_{Q,m_2} \rangle_{\mathrm{eq}}, \qquad (16)$$

$a_{Q,m,j}$, $\bar{a}_{Q,m,j}$ are right- and left-eigenvectors of $\mathbf{M}_Q$, and $\langle \hat{\sigma}_{-Q,m_1} \hat{\sigma}_{Q,m_2} \rangle_{\mathrm{eq}}$ are spatial Fourier transforms of static correlation functions. For the model described by the Hamiltonian in Eq. (1) we have

$$
\begin{aligned}
\langle \hat{\sigma}_{-Q,m} \hat{\sigma}_{Q,m'} \rangle_{\mathrm{eq}} &= \frac{1}{N} \sum_{l,l'} e^{-iQn(l-l')} \langle \sigma_{l,m} \sigma_{l',m'} \rangle \\
&= \frac{e^{iQn} v_n}{1 - e^{iQn} v_n} \frac{v_{m'-1}}{v_{m-1}} + \frac{e^{-iQn} v_n}{1 - e^{-iQn} v_n} \frac{v_{m-1}}{v_{m'-1}} \\
&\quad + \min\left\{ \frac{v_{m'-1}}{v_{m-1}}, \frac{v_{m-1}}{v_{m'-1}} \right\},
\end{aligned}
\qquad (17)
$$

where

$$v_{m-1} = \prod_{j=1}^{m-1} \tanh\left(\beta J_j\right), \qquad (18)$$

with $v_0 = 1$ and we have assumed all couplings to be ferromagnetic.

2.2. RESTRICTED DYNAMICS

Instead of considering the Glauber dynamics, we now assume a kind of transition associated to the motion of domain walls. The idea is similar but not identical to previous work by others in which either the domain wall motion is strongly suppressed at low temperatures [13], [14] or it is through a one–dimensional random walk [15]. This restricted dynamics leads to a transition probability for each spin which depends only on the configuration of one of its neighbors (the one on the left, for definiteness). The development is totally analogous to the one followed in the previous subsection, but in this case the transition probabilities take the form

$$w_{li}^{\text{restr}}(\sigma_{l,i}) = \alpha_i \left(1 - \tanh \beta \left(J_{i-1}\sigma_{l,i-1}\sigma_{l,i}\right)\right). \tag{19}$$

Introducing these transition probabilities into the calculation yields the following equation for the time evolution of the expectation values $q_{l,m}$

$$\begin{aligned}
\frac{dq_{l,m}}{dt} &= -\alpha_m \left[q_{l,m} - \langle \tanh \left(\beta J_{m-1}\sigma_{l,m-1}\right)\rangle\right] \tag{20}\\
&= -\alpha_m \left(q_{l,m} - \gamma_{m-1}q_{l,m-1}\right),
\end{aligned}$$

where $\gamma_n = \tanh\left(\beta J_n\right)$. By introducing Fourier transforms we again obtain a matrix differential equation similar to Eq. (9), but with a matrix given by

$$\mathbf{M}_Q^{\text{restr}} = \begin{pmatrix}
-\alpha_1 & 0 & 0 & \cdots & 0 & 0 & \alpha_1\gamma_n e^{iQn} \\
\alpha_2\gamma_1 & -\alpha_2 & 0 & \cdots & 0 & 0 & 0 \\
0 & \alpha_3\gamma_2 & -\alpha_3 & \cdots & 0 & 0 & 0 \\
\vdots & \vdots & \vdots & \ddots & \vdots & \vdots & \vdots \\
0 & 0 & 0 & \cdots & -\alpha_{n-2} & 0 & 0 \\
0 & 0 & 0 & \cdots & \alpha_{n-1}\gamma_{n-2} & -\alpha_{n-1} & 0 \\
0 & 0 & 0 & \cdots & 0 & \alpha_n\gamma_{n-1} & -\alpha_n
\end{pmatrix}. \tag{21}$$

¿From the eigenvectors and eigenvalues of this matrix we can once more obtain all the quantities of interest by using similar expressions to the ones derived for Glauber dynamics. It should be noted that since the transition probabilities for the restricted dynamics do not satisfy detailed balance, no equilibrium state is attained [5]. However, there is a steady state which is equivalent to an equilibrium state with an effective exchange constant equal to $\tilde{J}/2$. Therefore, with this equivalence, we can also obtain all the steady state properties from the results for the equilibrium state.

3.   RESULTS AND DISCUSSION

It should be clear that, irrespective of the dynamics considered, we have been able to obtain both the magnetization and the response function of the present model in closed form for arbitrary n. The actual computation of the eigenvectors and eigenvalues of the matrices $\mathbf{M}_Q$ and $\mathbf{M}_Q^{\text{restr}}$ may be laborious, but the problem is a standard one in linear algebra. The dynamical critical exponent z can be determined numerically from Eq. (12). For the Glauber dynamics the results are shown in Fig. 1 for cases where we have one or four different relaxation times. In both cases z was found to be equal to 3, satisfying the equation $z = 1 + |J_{\max}/J_{\min}|$, showing, as expected, that the nonuniversal behavior is not affected by the relaxation times. On the other hand, the similar results for the restricted dynamics presented in Fig. 2 show that $z = 2$ independently of the values of J's and α's.

In our previous work we have studied relaxation processes of the chains by looking at the imaginary part of the response function at zero momentum, $S_0(\omega)$. More specifically, we concentrated on the susceptibility $\chi(\omega) \equiv nk_BTS_0(\omega)/\sum_{m,m'}\langle\hat{\sigma}_{-Q,m}\hat{\sigma}_{Q,m'}\rangle_{\text{eq}}$ and its relationship to the Nagel scaling. However we have very recently pointed out a mistake in our former calculations [6] that, when corrected, does not lead to the neat connection between our kinetic Ising models and the so–called Nagel plots that we had found earlier. It should be recalled that in these plots, the abscissa is $(1 + W)\log_{10}(\omega/\omega_p)/W^2$ and the ordinate is $\log_{10}(\chi''(\omega)\omega_p/\omega\Delta\chi)/W$. Here, χ'' is the imaginary part of $\chi(\omega)$, W is the (logarithmic) full width at half maximum of χ'', ω_p the frequency corresponding to the peak in χ'', and $\Delta\chi = \chi(0) - \chi_\infty$ is the static susceptibility. The error resided in considering the linear rather than the logarithmic width in our computations. In this respect, a few more comments are in order. Only less than two years ago, what is now called the excess or the Nagel wing was found in the relaxation measurements of glass–forming ethanol [16]. It is believed that this exception to the Nagel scaling is due to multiple relaxation processes and shows up as a shoulder with some dispersion in the high frequency end. Preliminary calculations with the logarithmic width indicate that while the scaling is not so perfect there are some features reminiscent of the excess wing. Therefore two possibilities for further investigation immediately arise. The first one concerns a closer look at the (correct) Nagel plots arising in different kinetic Ising models on linear chains, including those we have studied before as well as others such as the random and aperiodic isotopic chain and the AB_2 and the A_2B_2 chains. The second one should evaluate the utility and appropriateness of the modified (using the linear width) Nagel plots both with respect to experimental data and in connection with other models. We are presently pursuing these two lines of research.

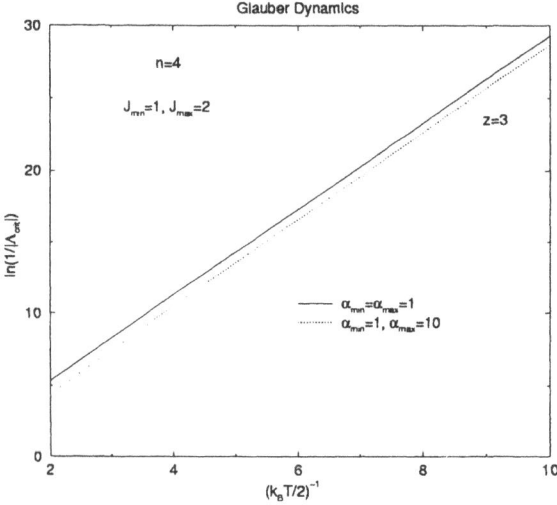

Figure 1.   Scaling relation (Glauber dynamics) for n=4 , J=1, 1+1/3,1+2/3, 2 and different values of α.

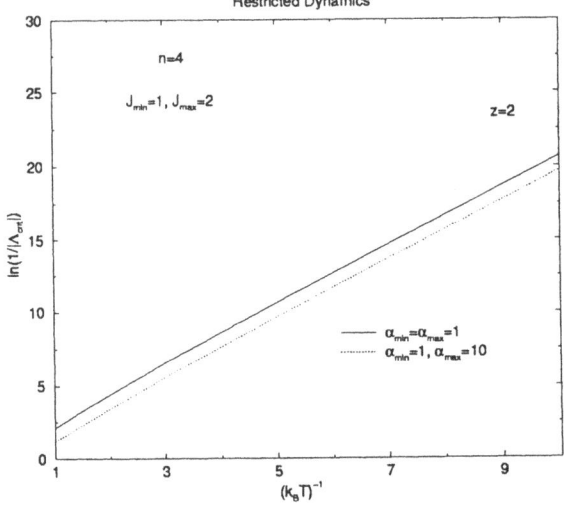

Figure 2.   Scaling relation (restricted dynamics) for n=4 , J=1, 1+1/3,1+2/3, 2 and different values of α.

Acknowledgments

Work partially financed by the Brazilian agencies CNPq, Finep and Fapesp and by DGAPA–UNAM under projects IN101100 and IN103100.

References

[1] A nice overview of the rich phenomenology of glassy systems and the importance of disorder for the underlying microscopic dynamics may be found in: *Disorder effects on relaxational processes. Glasses, Polymers, Proteins*, ed. R. Richert and A. Blumen. (Springer–Verlag, Berlin, 1994).

[2] P. K. Dixon, L. Wu and S. R. Nagel, B. D. Williams and J. P. Carini, *Phys. Rev. Lett.* **65** (1990) 1108; L. Wu, P. K. Dixon, S. Nagel, B. D. Williams, and J. P. Carini, *J. Non–Cryst. Solids* **131–133** (1991) 32; D. L. Lesley–Pelecky and N. O. Birge, *Phys. Rev. Lett.* **72** (1992) 1232; M. D. Ediger, C. A. Angell and S. R. Nagel, *J. Phys. Chem.* **100** (1996) 13200.

[3] L. L. Gonçalves, M. López de Haro, J. Tagüeña–Martínez and R. B. Stinchcombe, *Phys. Rev. Lett.* **84** (2000) 1507.

[4] L. L.Gonçalves, M. López de Haro, and J. Tagüeña–Martínez, *Braz. J. Phys.* **30** (2000) 731. Proceedings of the Ising Centennial Conference.

[5] L. L.Gonçalves, M. López de Haro, and J. Tagüeña–Martínez, *Phys. Rev.* **E63** (2001) 026114.

[6] L. L. Gonçalves, M. López de Haro, J. Tagüeña–Martínez and R. B. Stinchcombe, *Phys. Rev. Lett.* **88** (2002) 089901(E).

[7] R. J. Glauber, *J. Math. Phys.* **4** (1963) 294.

[8] M. López de Haro, J. Tagüeña–Martínez, B. Espinosa and L. L. Gonçalves, *J. Phys. A: Math. Gen.* **26** (1993) 6697; **29**(E) (1996) 7353; M. López de Haro, L. L. Gonçalves and J. Tagüeña–Martínez, *Mod. Phys. Lett.* **B10** (1996) 1441.

[9] J. C. Angles d'Auriac and R. Rammal, *J. Phys. A: Math. Gen.* **20** (1988) 763.

[10] L. L. Gonçalves and N. T. de Oliveira, *Can. J. Phys.* **63** (1985) 1215.

[11] M. Droz, J. Kamphorst Leal da Silva and A. Malaspinas, *Phys. Lett.* **A115** (1986) 448; M. Droz, J. Kamphorst Leal da Silva, A. Malaspinas and A. L. Stella, *J. Phys. A: Math. Gen.* **20** (1987) L387.

[12] B. I. Halperin and P. C. Hohenberg, *Phys. Rev.* **177** (1969) 952; P. C. Hohenberg and B. I. Halperin, *Rev. Mod. Phys.* **49** (1977) 435 .

[13] J. C. Kimball, *J. Stat. Phys.* **21** (1979) 289.

[14] U. Deker and F. Haake, *Z. Phys.* **B35** (1979) 281; F. Haake and K. Thol, *Z. Phys.* **B40** (1980) 219.

[15] J. L. Skinner, *J. Chem. Phys.* **79** (1983) 1955.

[16] R. Brand, P. Lunkenheimer, U. Schneider and A. Loidl, *Phys. Rev.* **B 62** (2000) 8878; U. Schneider, R. Brand, P. Lunkenheimer and A. Loidl, *Phys. Rev. Lett.* **84** (2000) 5560.

PRIMITIVE MODELS FOR THERMOTROPIC LIQUID–CRYSTALS

Niza Ibarra–Avalos, Alejandro Gil–Villegas
Instituto de Física, Universidad de Guanajuato,
León, Guanajuato, México.
gil@fisica.ugto.mx

Antonio Martinez–Richa
Facultad de Química, Universidad de Guanajuato,
Guanajuato, Guanajuato, México
richa@quijote.ugto.mx

Abstract   The phase behavior and structural properties of thermotropic liquid crystals can be studied using primitive models that must satisfy two requirements: (a) the molecular model has the basic elements that determine the most relevant features observed meso and macroscopically, and (b) a system formed by these models can be systematically studied through computer simulation and molecular-based theories. In this paper we review some advances in the development of molecular thermodynamic theories using primitive models characterized by computer simulation and information of real substances provided by energy minimization with the PM3 semiempirical method.

Keywords: Liquid Crystals, perturbation theory, computer simulation, polar interactions.

1.   INTRODUCTION

Liquid crystal phases (LC) are condensed phases with intermediate orientational and/or positional order, exhibiting lower symmetry than a liquid phase but higher than a solid or plastic crystal phases. The degrees of freedom of LC molecules can correlate liquid-like or solid-like, inducing anisotropic phases. As for example, nematic phases have orientational ordering in one (uniaxial) or two directions (biaxial) but the centers of gravity of the constituent molecules have a short-range order as in an isotropic phase. Organic compounds that are mesogenic have

Developments in Mathematical and Experimental Physics,
Volume B: Statistical Physics and Beyond
Edited by Macias *et al.*, Kluwer Academic/Plenum Publishers, 2003

227

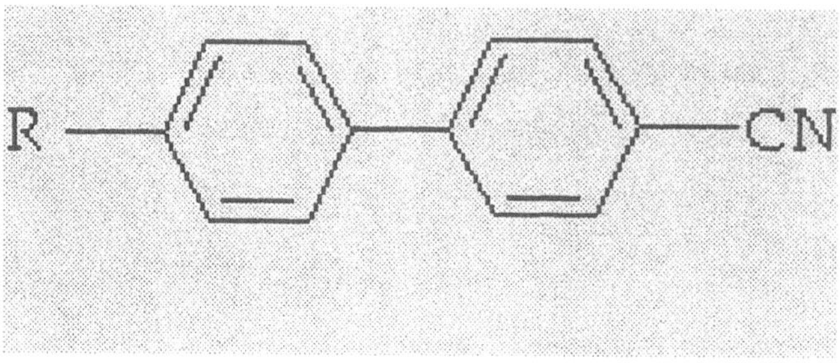

Figure 1.   Alkylcyanobiphenyl, synthesized by Gray and coworkers in 1972 [1]

rod-like or disk-like geometries, and the anisotropic shape is the dominant interaction that produces LC phases. A typical example of LC molecule is given in figure 1, that corresponds to an alkylcyanobiphenyl. These organic compounds were synthesized by Gray and coworkers in 1972 at the University of Hull [1], and they were the first commercially nematic liquid crystals used in display devices. These compounds are chemically and photochemically stable. The positive dielectric and reasonably high optical anisotropy have been associated with the aromatic core in conjugation with the cyano group. Transition LC temperatures, phase stability and other physical properties of liquid crystals with flexible chains depend upon alkyl chain length. The nematic phase stability and melting points for these substances are influenced by overall molecular flexibility. Short terminal chains involve rigid structures with high melting points. As the chain length is incremented, melting points are lowered because of increasing flexibility, but for very long chains (i.e., higher than C9) van der Waals attraction forces are important and then the melting points are again increased. It is this delicate balance between structure and interactions that gives the complexity in the behavior of LC substances, and, consequently, their very rich phase diagrams.

In order to model LC substances, we require to determine which are the basic molecular features that we need to take into account for a proper modelling. Although the detailed atomistic description of LC molecules could be used for obtaining information about mesoscopic phases, this route is computationally very demanding. Over the years, the use of *primitive* molecular models has been very useful to obtain qualitative predictions. By primitive model (PM) we mean a crude representation of actual LC molecules but containing the basic microscopic information, such as molecular shape and intermolecular forces. Computer simulation can then be used for these simplified models in order to understand the interplay of the different molecular parameters, whereas phenomenological and microscopic theories can be applied to predict phase diagrams.

In this paper we present advances in our understanding of real mesogenic substances, using a combination of chemical and physical methods that enhances the application of primitive models to real LC substances.

2. NON–POLAR LC MOLECULES

The role of van der Waals forces has been studied with PM models consisting in convex bodies surrounded by a square–well potential [2, 3, 4]. As demonstrated by Gelbart and Gelbart [5], molecular models comprised of an anisotropic core plus isotropic attractive interactions give the basic orientational behavior in nematic, and the findings of the mentioned studies are that the phase diagram for these molecules consists of an anisotropic, a nematic and smectic phase. The relevance of these PM models is that they can be modelled through molecular-based thermodynamic theories, such as the Barker and Henderson perturbation theory [6] extended to anisotropic systems. Williamson and Guevara [7] have shown that a proper selection of the square–well range parameter, reproduces very accurately the liquid–vapor coexistence curve obtained by computer simulation for other PM models, such as ellipsoids interacting with the Gay-Berne potential [8].

3. POLAR LC MOLECULES

A PM model for 4–n–octyl–4'–cyanobiphenyl (8CB) has been proposed recently by van Duijneveldt *et al.* [9]. This model consists of a hard spherocylinder, a terminal point dipole and an ideal flexible tail. The hard spherocylinder models the rigid part of the molecule, given by the aromatic core, and has an aspect ratio $L/D = 5$, where L is the length of the cylinder and D its diameter. The dipole is placed in the center of an end cap, aligned parallel to the main molecular axis, and the tail is attached at the opposite end of the spherocylinder, consisting of five segments of length D, with a bond angle between successive segments fixed at 109.5°.

The ideal character of the tail consists in that there are no excluded volume effects between tails, only between the end point of each segment and the spherocylinder core. Van der Waals attractions are absent in this model. As reported in reference [9], this PM model presents isotropic, nematic and smectic–A phases, in the same sequence as 8CB. Studying the effect of the different molecular components of the model, it results that the stability of the nematic phase is affected in opposite ways by the presence of the dipole and the tail. Whereas the terminal dipole enhances the range of stability of the nematic phase [10], the tail suppress it [11]. Combining the effects of the dipole and the tail, it results that a narrow range of stability of the nematic phase is observed. This opposing effect have also a marked influence on the layered phases.

4.   THERMODYNAMIC MODELLING

The theoretical prediction of LC phase diagrams using PM models can be achieved through a proper thermodynamic potential, derived from a phenomenological approach or from a molecular–based theory. Here we summarize recent developments following the second approach.

In order to have a proper description of the thermodynamic effects arising from the different molecular elements, the Helmholtz free energy for a LC substance comprised of N molecules can be given as the sum of the different contributions produced by those molecular elements. In the case of the non–polar PM models discussed in the previous section, the free energy can be expressed as

$$\frac{A}{NkT} = \frac{A_{ideal}}{NkT} + \frac{A_\Omega}{NkT} + \frac{A_0}{NkT} + \frac{A_{vdW}}{NkT} \qquad (1)$$

where k is the Boltzmann's constant, T is the temperature, A_{ideal} is the ideal contribution to the free energy, A_Ω is an orientational free energy — Onsager's formula [12] —, A_0 is the Helmholtz free energy of the hard–core body, and A_{vdW} is the contribution to the free energy from the van der Waals forces.

The calculation of A_0 can be done using a scaling procedure proposed by Parsons [13], that basically consists in giving the non-spherical hard body properties in terms of the hard–spheres fluid properties. It is not clear why this approach works so well for convex bodies with long elongations ($L/D \geq 5$), although there are evidences that it can be formally derived from kinetic theory [14]. The Parsons approach is used also for the evaluation of A_{vdW}, in combination with the Barker and Henderson approach [6]. The developed theory can be applied to the PM models already presented, and has been used for the description of the phase diagram of a typical LC substance, such as p–azoxyanisole [4, 15] The thermodynamic modelling of real LC substances, as p–azoxyanisole, can be substantially improved if a more realistic estimation of the hard-body

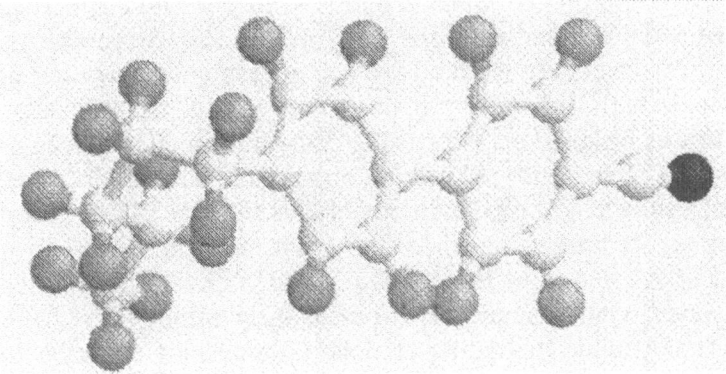

Figure 2. Minimum energy geometry for n–pentyl–cyanobiphenyl obtained by PM3 method. Representation is as follows: black for Nitrogen atom, gray for Hydrogen atoms, and white for Carbon atoms

geometric parameters is taken. This can be accomplished by energy minimizations using PM3 semiempirical calculations, which uses experimental values of the angles and distances between atoms in order to get the molecular structure. In reference [15] we have reported the combined use of this technique with the perturbation theory briefly outlined here. Figure 2 shows the molecular structure of another LC substance (polar in this case), n–pentyl–cyanobiphenyl, obtained by the PM3 method. Notice that due to the rigidity of the aromatic rings, a more realistic hard-core body should be not a revolution body but a body with minor axes of different length. The modelling gives also information about the position and orientation of the dipole, a valuable information also for more accurate computer simulation studies of the PM model. The thermodynamic modelling of polar LC substances is more challenging, since polar interactions give rise to a complexity in the stability of LC phases, as has been already established in a series of Monte Carlo studies for dipolar PM models, such as hard spherocylinders with point dipoles [9, 10, 16, 17] and dipolar Gay Berne sytems [18, 19, 20]. Different theories for these models have predicted opposite effects. For example, Vega and Lago [21], and Vanakaras and Photinos [22] predicted that elongated hard–core molecules with longitudinal central dipoles should stabilize the nematic phase with respect to the isotropic phase, in contradiction with computer simulation results by McGrother *et al.* [10, 16]. These authors found that the polar molecules exhibit short-ranged antiferroelectricity, and that by the pairing of molecules the effective aspect ratio of the resultant cluster enhances the stability of the isotropic phase. Williamson and del Río [23] proposed a perturbation theory for non–spherical polar molecules and by including three body effects they were able to show that the isotropic phase is stabilized. A proper understanding of these systems came with the theory proposed by Emelyanenko and Osipov [24]. These authors included an associating contribution to the free energy in order to take into account the pairing observed in the simulated systems. For small values of the dipolar moment the nematic phase is stabilized but there is a critical value where the effect is reversed and for dipolar moments above this value the nematic phase is distabilized. The theory of Emelyanenko and Osipov [24] is particularly enlightening in showing that thermodynamic modelling of polar interactions must be handled with caution, even for simple polar fluids. Their theory proposes a polar free energy given by four terms, the usual dipole–dipole contribution, plus the contributions of the dimer (i.e, dimer formed by the pairing of dipoles) formation, and the monomer–dimer and dimer–dimer interactions. Notice that by neglecting the associating free energy arising from the dimer formation, the inversion mechanism for the isotropic-nematic range of stability *can not be obtained*, a clear indication that a proper thermodynamic modelling of polar substances is incomplete if the associating mechanism is not included. We think that the combined use of Emelyanenko and Osipov [24] methodology, together with perturbation theories and molecular characterization by PM3 calculations, would provide realistic molecular thermodynamic theories for polar LC substances, like the alkylcyanobiphenyls.

5.   QUADRUPOLAR INTERACTIONS

Computer simulations give insights into complex behavior even when a very simplistic model is used. Figure 3 shows a typical low-temperature configuration observed for axial point quadrupolar hard spheres, obtained by Monte Carlo simulation in the NVT ensemble, for 256 particles. The spheres are depicted with three bands, in order to give information about the quadrupolar axis and to clarify that we are modelling the configuration of basically four charges aligned with the axis. This configuration exhibits an antiferroelectric–like ordering of the quadrupolar axis, which resembles a pattern of "v" structures. This pattern has been observed experimentally in real quadrupolar LC substances [25]. In spite of the complexity of the organic substance studied, it seems that the main molecular effect observed experimentally can be explained by the quadrupolar interaction.

Figure 3. Monte Carlo configuration of axial quadrupolar hard spheres of diameter σ and quadrupolar moment $Q*^2 = 3.025$, where $Q*^2 = Q^2/kT\sigma^5$, for a density $\rho* = \rho\sigma^3 = 0.90$. The simulation was performed with 256 particles.

6.   CONCLUSIONS

Primitive models can be implemented for the study of real LC substances when combined with computer simulations, minimum energy calculations as the PM3 method, and molecular–based theories. Although non–polar LC substances can be described using these approaches, the case of polar substances requires more development but there are already theoretical approaches that have enlightened the route to follow: by including a more detailed description of the molecular effects induced by polar forces, such as the associating mechanism present in dipolar molecules, the thermodynamic stability of LC phases can be properly described.

Acknowledgments

N.I.A. acknowledges support from CONACYT (México) for graduate studies. We also acknowledge funding from CONACYT, grant 33523–E.

References

[1] G. W. Gray, *Thermotropic Liquid Crystals*, (Wiley, USA, 1987).

[2] E. de Miguel and M. P. Allen., *Mol. Phys.* **76** (1992) 1275.

[3] D. C. Williamson and F. del Río, *J. Chem. Phys.* **109** (1998) 4675.

[4] E. García, D. C. Williamson and A. Martínez–Richa, Mol. Phys. **98** (2000) 179.

[5] W. M. Gelbart and A. Gelbart, *Mol. Phys.* **33** (1977) 1387.

[6] J. A. Barker and D. Henderson, *Rev. Mod. Phys.* **48** (1976) 587.

[7] D. C. Williamson and Y. Guevara, *J. Phys. Chem. B* **103** (1999) 7522.

[8] E. de Miguel, L. F. Rull, M. K. Chalam and K. E. Gubbins, *Mol. Phys.* **74** (1991) 405.

[9] J. S. van Duijneveldt, A. Gil–Villegas, G. Jackson and M. P. Allen, *J. Chem. Phys.* **112** (2000) 9092.

[10] S. C. McGrother, A. Gil–Villegas and G. Jackson, *J. Phys. Cond. Matter* **8** (1996) 9649.

[11] J. S. van Duijneveldt and M. P. Allen, *Mol. Phys.* **92** (1997) 855.

[12] L. Onsager, *Ann, Acad. Sci.* **51** (1949) 627.

[13] J. D. Parsons, *Phys. Rev. A* **19** (1979) 1225.

[14] A. Gil–Villegas and F. del Río, *in preparation*.

[15] E. García–Sánchez, A. Martínez–Richa, J. A. Villegas–Gasca, L. H., Mendoza–Huizar and A. Gil–Villegas, *J. Phys. Chem. B* (2001) *submitted*.

[16] S. C. McGrother, A. Gil–Villegas and G. Jackson, *Mol. Phys.* **95** (1998) 657.

[17] A. Gil–Villegas, S. C. McGrother and G. Jackson, *Chem. Phys. Lett.* **269** (1997) 441.

[18] K. Satoh, S. Mita and S. Kondo, *Chem. Phys. Lett.* **255** (1996) 99.

[19] R. Berardi, S. Orlandi and C. Zannoni, *Chem. Phys. Lett.* **261** (1996) 357.

[20] M. Houssa, A. Oulid and L. F. Rull, *Mol. Phys.* **94** (1998) 439.

[21] C. Vega and S. Lago, *J. Chem. Phys.* **100** (1994) 6727.

[22] A. G. Vanakaras and D. J. Photinos, *Mol. Phys.* **85** (1985) 1089.

[23] D. C. Williamson and F. del Río, *J. Chem. Phys.* **107** (1997) 9549.

[24] A. V. Emelyanenko and M. A. Osipov, *Cristallogr. Rep.* **45** (2000) 501.

[25] M. Villanueva–García, J. Robles and A. Martínez–Richa, *Comp. Mat. Science* **22** (2001) 300.

MODELLING THERMODYNAMIC PROPERTIES OF FLUIDS WITH DISCRETE POTENTIALS

Ana Laura Benavides
Instituto de Física, Universidad de Guanajuato,
León, Guanajuato, México
alb@fisica.ugto.mx

Alejandro Gil–Villegas
Instituto de Física, Universidad de Guanajuato,
León, Guanajuato, México
gil@fisica.ugto.mx

Abstract   Thermodynamic properties of fluids can be predicted by discretizing real intermolecular interactions and applying a perturbation theory developed for these discrete potentials. The theory requires to know the properties of the building–block fluids, i.e., square–well (SW) and square–shoulder (SS) fluids of variable range. We present applications of this method to fluids formed by chain molecules and calculation of exact second virial coefficients.

Keywords: Discrete potentials, perturbation theory, virial coefficients, chain molecules.

1.   INTRODUCTION

Perturbation theory (PT) in Statistical Mechanics is a powerful method that enable us to predict thermodynamic and structural properties of fluid substances. The hard–sphere fluid (HS) is an excellent reference system in the PT methodology, since the short–range repulsive part of real interatomic potentials, which dominates the correlations between atoms, can be represented by the discrete hard-sphere repulsion [1, 2]. In the other hand, the van der Waals attractive forces are taken as perturbations of the HS system. The success of the PT approaches in modelling real substances relays on the fast convergence of the perturbation expansions.

Developments in Mathematical and Experimental Physics,
Volume B: Statistical Physics and Beyond
Edited by Macias *et al.*, Kluwer Academic/Plenum Publishers, 2003

Furthermore, other interactions can be taken into account in perturbation expansions, such as polar forces and short and long range repulsive forces. Arbitrary discrete potentials can be built via a sequence of square–well and square-shoulder potentials, and their properties can be obtained by discontinuous molecular dynamics [3] and perturbation theory (DPT) [4]. The DPT approach provides a framework for studying primitive models of fluids composed by spherical particles interacting via a discrete potential. The free energy of any hard–core discrete potential formed by a sequence of square–well and square–shoulder steps can be essentially given in terms of the properties of square–well fluids. This simple result has been applied to several potential models used in different contexts, and the theory works very well in almost all the cases, for pure fluids and mixtures [4, 5]. Additionally, with the DPT method is possible to obtain the equation of state for real potentials whose mathematical form is unknown.

In this paper we continue the applications of the DPT approach in the context of low–density fluids and chain molecule fluids.

2.   THEORY

We consider a system of N spherical particles of diameter σ_0 contained in a volume V, which are interacting with an arbitrary discrete potential u_d given by:

$$u_d(r) = u_{HS}(r, \sigma_0) + \sum_{i}^{m} \phi_i(r) \tag{1}$$

where $u_{HS}(r, \sigma_0)$ is the hard–spheres repulsive contribution, m is the number of steps, and the potential ϕ_i is defined by

$$\phi_i(r) = \begin{cases} \epsilon_i & \text{if } \lambda_{i-1}\sigma_0 < r < \lambda_i\sigma_0 \\ 0 & \text{otherwise} \end{cases} \tag{2}$$

where ϵ_i could be positive or negative, and $\lambda_0 = 1$. The set of steps ϕ_i define the shape of a general potential u_d. By a proper election of several repulsive and attractive steps it is possible to obtain approximated versions of continuous model potentials, such as Lennard–Jones [6] or more complex interactions in the case of colloidal particles and polymers. See Figures 1 to 3.

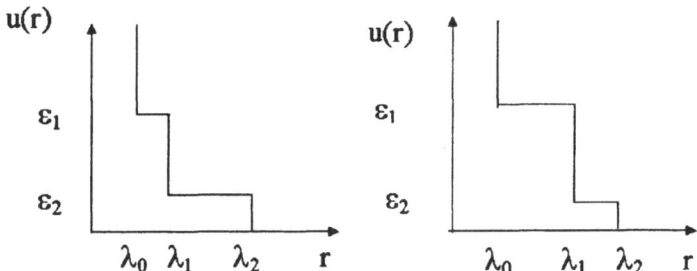

Figure 1. Examples of discrete potentials with the same energy depths and different ranges. These potential models can be used to study colloidal behavior.

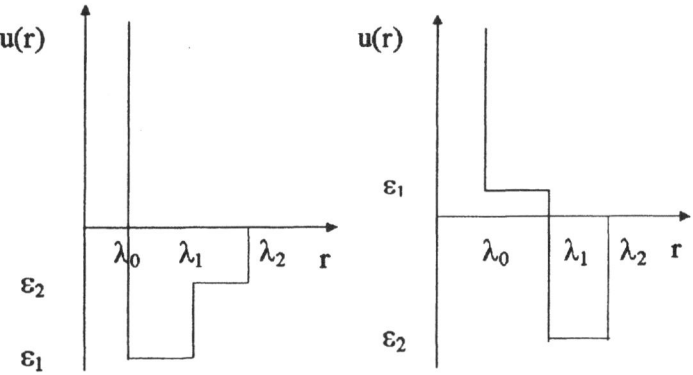

Figure 2. Examples of discrete potentials with the same ranges and different energy depths. The potential depicted in the right hand corresponds to a model used for supercooled water [10]

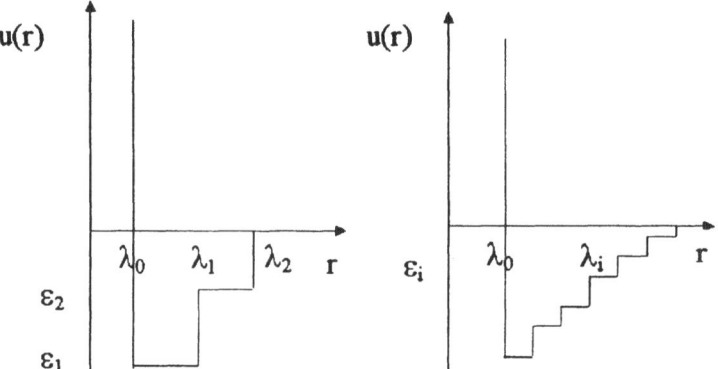

Figure 3. Examples of two discrete potentials with variable values of steps, energy depths and ranges. The potential depicted in the right hand corresponds to a model used for vibrating square–well chains [11]

The Helmholtz free energy is obtained from a high–temperature expansion (HTE) to second order [1]:

$$\frac{A}{NkT} = \frac{A^{HS}}{NkT} + \beta \sum_{i=1}^{m} \left[a_1^S(\eta, \lambda_i, \epsilon_i) - a_1^S(\eta, \lambda_{i-1}, \epsilon_i) \right]$$

$$+ \beta^2 \sum_{i=1}^{m} \left[a_2^S(\eta, \lambda_i, \epsilon_i) - a_2^S(\eta, \lambda_{i-1}, \epsilon_i) \right] \qquad (3)$$

where A_{HS} is the free energy for the hard–spheres reference fluid, a_1^S and a_2^S are the first and second order perturbation terms for a square–well ($\epsilon_i < 0$) or square–shoulder ($\epsilon_i > 0$) fluid, and $\beta = 1/kT$, where k is the Boltzmann's constant and T is the temperature. See reference [4] for more details.

As we have shown, the thermodynamic properties of a general discrete–potential system are given by the corresponding properties of elementary systems interacting via the SW and SS potentials. The thermodynamic properties for both models are essentially given by the same Helmholtz free energy HTE expression, once that we have taken into account the sign of the respective depth/height energy parameter ϵ_i.

In the case of the square–shoulder fluid we can construct an EOS using the SW results:

$$a_1^{SS}(\eta, \lambda) = -a_1^{SW}(\eta, \lambda) \qquad (4)$$

and

$$a_2^{SS}(\eta, \lambda) = a_2^{SW}(\eta, \lambda) \qquad (5)$$

where λ is the range of the potential.

For both systems, the free energy can be expressed as

$$\frac{A}{NkT} = \frac{A_{HS}}{NkT} \pm \beta a_1^{SW} + \beta^2 a_2^{SW} \qquad (6)$$

where the positive and negative signs correspond to a square–well and square–shoulder systems, respectively. In order to study the SS properties we require a SW EOS. We have used the SW EOS of reference [7] for $1.25 < \lambda < 2.0$ and the long range square-well EOS for $\lambda > 2.0$ [8] since they describe very well the SW properties in a wide range of temperatures and densities. These SW EOS provide analytical expressions for a_1 and a_2; the second–order term is more accurate than the corresponding local compressibility approximation (LCA) or macroscopic compressibility approximation (MCA) values proposed by Barker and Henderson [9]. We have used these EOS to illustrate the method, although other theories for SW fluids can be used.

3.   CHAIN MOLECULES

An important application of the DPT approach is the prediction of thermodynamic properties of fluids formed by chain molecules, when combined with molecular–based equations of state, as for example the Statistical Associating Fluid Theory (SAFT) [12, 13]. Both approaches can be particularly useful when the real intermolecular potential has an unknown mathematical form and a discretized potential can be used instead. The SAFT approach requires to know the Helmholtz free energy of the segments that form the chain molecules and the contact value of the corresponding radial distribution function. Assuming a discrete potential model, Eq.(1), the Helmholtz free energy is given by Eq.(2), whereas the contact value of the radial distribution function can be obtained from perturbation theory, using a self–consistent method for the calculation of the pressure, as explained in reference [14]. The expression obtained following this method is

$$g(\sigma_0^+) = g_{HS}(\sigma_0^+) + \beta g_1 \qquad (7)$$

where $g_{HS}(\sigma_0^+)$ is the contact value of the hard–spheres radial distribution function, and g_1 is the first perturbation term, given by

$$g_1 = \frac{1}{4}\frac{\partial a_1}{\partial \eta} + \sum_{i=1}^{m} \lambda_i^3 \sigma_0^3 \left[\phi(\lambda_i^+ \sigma_0) - \phi(\lambda_i^- \sigma_0)\right] g_{HS}(\lambda_i \sigma_0) \qquad (8)$$

and

$$a_1 = \sum_{i=1}^{m} \left[a_1^S(\eta, \lambda_i, \epsilon_i) - a_1^S(\eta, \lambda_{i-1}, \epsilon_i) \right] \tag{9}$$

where λ_i^+ and λ_i^- indicate the right and left limits, respectively. The Helmholtz free energy A^{ch} for a fluid formed by M chain molecules, being n the number of spherical segments per chain, is given by

$$\frac{A^{ch}}{MkT} = \frac{A^{ideal}}{MkT} + n\frac{A}{NkT} - (n-1)\ln y(\sigma_0) \tag{10}$$

where $y(\sigma_0) = g(\sigma_0^+) \exp(\beta u_d(\sigma_0^+))$, and A^{ideal} is the ideal gas contribution to the Helmholtz free energy.

The SAFT approach has been used to model asphaltene precipitation using very simplified intermolecular potentials [15, 16]. A more detailed description of the actual asphaltene–asphaltene, asphaltene–resin and resin–resin intermolecular potentials can be obtained through the use of discrete models. The DPT approach can be incorporated into the SAFT methodology, as we have outlined here (Eq.(10)), in order to predict the phase diagram of these substances.

4.   SECOND VIRIAL COEFFICIENT

An exact result that is obtained with the use of a discrete potential, Eq.(1), is the second virial coefficient. A straightforward calculation of this coefficient for a gas of particles interacting with the discrete potential (1) gives the following result

$$B_2 = \frac{2\pi}{3} \sum_{i=0}^{m} \lambda_i^3 \sigma_0^3 \left[e^{-\beta u_d(\lambda_i^+ \sigma_0)} - e^{-\beta u_d(\lambda_i^- \sigma_0)} \right] \tag{11}$$

where λ_i^+ and λ_i^- indicate the right and left limits, respectively. This result could be very useful not only for diluted gases but also in liquid crystals theories based on the Onsager approach, as we will show in a future communication.

5.   CONCLUSIONS

The DPT approach provides a framework for studying fluids composed by spherical particles interacting via a discrete potential, pure fluids and mixtures. As we have briefly outlined here, the approach has also an important application in the prediction of thermodynamic properties of more complex fluids, such as chain molecules systems and liquid crystals.

References

[1] J. A. Barker and D. Henderson, *Rev. Mod. Phys.* **48** (1976) 587.

[2] W. C. K.Poon and P. N. Pusey, in: *Observation, Prediction and Simulation of Phase Transitions in Complex Fluids*, NATO ASI series C, Vol. 460 (Kluwer Academic Publishers, 1995).

[3] G. A. Chapela, S. E. Martinez–Casas and J. Alejandre, *Molec. Phys.* **53** (1984) 139.

[4] A. L. Benavides and A. Gil–Villegas, *Mol. Phys.*, **97** (1999) 1225.

[5] A. Vidales, A. L. Benavides and A. Gil–Villegas, *Mol. Phys.*, **99** (2001) 703.

[6] G. A. Chapela, L. E. Scriven and H. T. Davis, *J. Chem. Phys.* **91** (1989) 4307.

[7] A. Gil-Villegas, F. del Rio and A. L. Benavides, *Fluid Phase Equil.* **119** (1996) 97.

[8] A. L. Benavides and F. Del Rio, *Molec. Phys.* **68** (1989) 983.

[9] J. A. Barker and D. Henderson, *J. Chem. Phys.* **47** (1967) 2856.

[10] G. Franzese, G. Malescio, A. Skibinsky, S. V. Buldyrev and H.E. Stanley, *Nature* **409** (2001) 692.

[11] L.G. Hu, H. Rangwalla, J. Y. Cui and J. R. Elliot, *J. Chem. Phys.* **111** (1999) 1293.

[12] W. G. Chapman, K. E. Gubbins, G. Jackson and M. Radosz, *Fluid Phase Equil.* **52** (1989) 31.

[13] W. G. Chapman, K. E. Gubbins, G. Jackson and M. Radosz, *Ind. Eng. Chem. Res.* **29** (1990) 1709.

[14] A. Gil–Villegas, A. Galindo, P. J. Whitehead, J. S. Mills, G. Jackson, and A. N. Burgess, *J. Chem. Phys.* **106** (1997) 4168.

[15] J. Wu, J. M. Prausnitz, and A. Firoozabadi, *AIChE J.* **44** (1998) 1188.

[16] E. Buenrostro-González, A. Gil–Villegas, and C. Lira-Galeana, *Thermodynamics of Asphaltene precipitation using the SAFT-VR approach*, to be published in the Proceedings of the AIChE Spring Meeting 2002, U.S.A.

THERMODYNAMIC PROPERTIES OF FLUIDS DETERMINED FROM THE SPEED OF SOUND

Andrés F. Estrada–Alexanders

Departamento de Física, Universidad Autónoma Metropolitana
Av. San Rafael Atlixco 186, Col. Vicentina, Iztapalapa, México, D.F. CP 09340

afea@mixtli.uam.mx

Daimler Justo

Departamento de Física, Universidad Autónoma Metropolitana
Av. San Rafael Atlixco 186, Col. Vicentina, Iztapalapa, México, D.F. CP 09340

Abstract   In this paper we discuss a new approach to derived thermophysical properties of fluids from essentially speed of sound data. The speed of sound is linked to both the compression factor and the heat capacity at constant volume by two partial differential equations. Previously, this pair of equations has been solved as an initial value problem in which a numerical solution was obtained in the region where speed of sound data is available. Here, we transformed that pair of partial differential equations into a single quasi–linear partial differential equation which was solved subject to boundary conditions in a numerical way as well. This numerical solution also gives the thermodynamic properties in the region where speed of sound data is available. Finally, we compared the results of both ways of deriving thermophysical properties against an empirical equation of state for argon in the gaseous phase.

Keywords: Speed of sound, heat capacities, compression factor.

1.   INTRODUCTION

Accurate determinations of thermodynamic properties of fluids are still a matter of great interest. Some of these properties as heat capacities and pVT data can be measured directly and some other are derived from the quantities measured by means of thermodynamic relationships. The speed of sound is a property which can be measured with an outstanding precision (better than few hundreds of parts per million) [1]

Developments in Mathematical and Experimental Physics,
Volume B: Statistical Physics and Beyond
Edited by Macias *et al.*, Kluwer Academic/Plenum Publishers, 2003

243

and very fast by means of the current experimental techniques. In contrast, heat capacities are usually determined with a modest precision (about 0.5 per cent) and the measurements are time–consuming. This motivated the use of speed–of–sound measurements as an alternative way to obtain information about heat capacities. So far, several methods have been introduced in the literature with the aim of exploiting the precision of speed-of-sound measurements, [2–7] but only one [8] did not make approximations. In [8], the partial differential equations which link the speed of sound with the thermodynamic properties were integrated numerically by setting the problem as one of initial value (IV) and then, implementing a predictor–corrector integration algorithm. However, as the set of equations to be solved are no–linear, small errors due to the truncation of the Taylor's series used during the IV integration, propagated and introduced small ripples in the heat capacities derived from the integration. Besides, these ripples were enhanced by the use of polynomial interpolation of the speed of sound data at states where direct measurement were not available. Later, we found evidence that pVT determinations shown ripples, but in a very small scale. Therefore, we looked for another method to solve the same problem but avoiding the ripples–formation. We suggested to change from an initial value problem to a boundary condition problem (BC). This paper shows how the BC method is implemented and the results obtained from it; finally, we present a comparison between these two methods and give some conclusions.

2. THEORY

The speed of sound u for a homogeneous fluid at pressure p and at mass–density ρ is defined by:

$$u^2 = (\partial p/\partial \rho)_S, \qquad (1)$$

where S is the entropy. Introducing the compression factor $Z = p/RT\rho_n$, as a function of temperature T and molar density $\rho_n = \rho/M$ (with R is universal gas constant and M the molar mass) and changing the direction of the partial derivative from isentropes to isotherms, (1) is rewritten

$$u^2 M/RT = Z + \rho_n(\partial Z/\partial \rho_n)_T + (R/C_{V,\mathrm{m}})[Z + T(\partial Z/\partial T)_{\rho_n}]^2, \quad (2)$$

where $C_{V,\mathrm{m}}$ is the molar heat–capacity at constant volume. Equation (2) may be solved together with the following relationship:

$$\rho_n(\partial C_{V,\mathrm{m}}/\partial \rho_n)_T = -R[2T(\partial Z/\partial T)_{\rho_n} + T^2(\partial^2 Z/\partial T^2)_{\rho_n}]. \qquad (3)$$

Equations (2) and (3) are a coupled partial differential equations system which has been solved by an integration method relying on initial values for Z and $(\partial Z/\partial T)$ at the initial temperature. [8] Instead of working with (2) and (3) in an IV approach, we combined them into a single quasi-linear partial differential equation on Z. We solved (2) for $C_{V,m}$ and then took its partial derivative with respect to ρ_n in order to eliminate $C_{V,m}$ from (3); so, we found the next expression for Z:

$$A(\partial^2 Z/\partial \rho_n^2)_T + B(\partial^2 Z/\partial \rho_n \partial T) + C(\partial^2 Z/\partial T^2)_{\rho_n} = G. \qquad (4)$$

In (4), the coefficients A, B, C and G, are actually no-linear functions of $T, \rho_n, Z, (\partial Z/\partial \rho_n), (\partial Z/\partial T)$ and u,

$$A = (Ta)^2 = T^2 \left[F - Z - \rho_n \left(\frac{\partial Z}{\partial \rho_n} \right) \right]^2, \qquad (5)$$

$$B = 2T\rho_n \left[Z + \left(\frac{\partial Z}{\partial T} \right) \right] \left[F - Z - \rho_n \left(\frac{\partial Z}{\partial \rho_n} \right) \right], \qquad (6)$$

$$C = (\rho_n c)^2 = \rho_n^2 \left[Z + \left(\frac{\partial Z}{\partial T} \right) \right]^2, \qquad (7)$$

$$G = \rho_n c^2 \left[\left(\frac{\partial F}{\partial \rho_n} \right) - 2 \left(\frac{\partial Z}{\partial \rho_n} \right) \right]$$
$$-2a \left[T(F - Z) \left(\frac{\partial Z}{\partial T} \right) + \rho_n Z \left(\frac{\partial Z}{\partial \rho_n} \right) \right], \qquad (8)$$

where $F = u^2 M/RT$ is a dimensionless speed of sound. It is interesting to note that (4) is a parabolic equation in the whole (T, ρ_n) plane; i.e. $B^2 - 4AC = 0$ with $A > 0$ and $C > 0$. An analytical solution of (4) is not available but it is possible to get a numerical solution subjected to boundary conditions in the (T, ρ_n) region (either the vapor or the liquid phase) where u is known. Besides, $C_{V,m}$ and the molar heat capacities at constant pressure $C_{p,m}$ can be obtained from Z and F by $C_{V,m} = c^2 R/a$, and $C_{p,m} = c^2 RF/(a[F - a])$.

3. NUMERICAL INTEGRATION

Before going further, it is convenient to introduce the reduced variables $\tau = T/T^c$ and $\delta = \rho_n/\rho_n^c$ (the superscript c denotes a critical constant). In terms of these variables, (4) is written as:

$$a^2[\tau^2(\partial^2 Z/\partial \tau^2) + 2\tau(\partial Z/\partial \tau)] + 2ac[\delta\tau(\partial^2 Z/\partial \delta \partial \tau) + \delta(\partial Z/\partial \delta)] +$$
$$c^2[\delta^2(\partial^2 Z/\partial \delta^2) + 2\delta(\partial Z/\partial \delta) - \delta(\partial F/\partial \delta)] = 0. \qquad (9)$$

We approximated in (9) all partial derivatives by finite difference equations by means of a central-difference formula. Then, the region where (9) would be solved was discretized and represented by a regular mesh with size $\Delta\tau$ and $\Delta\delta$, and τ and δ were determined by a pair of positive–integer indices j and i, respectively; i.e. $\tau = \tau_j = j\Delta\tau$ and $\delta = \delta_i = i\Delta\delta$. With this approach, at any point of the mesh the values of Z and F were denoted by two subscripts j and i; so $Z(\tau, \delta) = Z(\tau_j, \delta_i) = Z_{j,i}$. The mesh's location in the (τ, δ) plane is given easily by choosing a minimum and a maximum value to each index, eg. j may vary between $j_{min} = j_0$ and $j_{max} = J$. Selecting properly j_0, we can studied either the supercritical region ($\tau > 1$) or the subcritical region ($\tau < 1$) and with the selection of i_0, the region covered can start from the zero–pressure limit ($\delta = 0$) or at a finite density. All partial derivatives of Z and F and the functions a and c were discretized and we replaced them in (9). This procedure transforms the no–linear partial differential equation on Z into the following set of algebraic cubic equations

$$Z_{j,i}^3 + b_2 Z_{j,i}^2 + b_1 Z_{j,i} + b_0 = 0, \tag{10}$$

in which each b_l (with $l = 0, 1, 2$) is no-linear function of the eight nearest neighbors of $Z_{j,i}$ and also of $F_{j,i+1}, F_{j,i}$ and $F_{j,i-1}$. The explicit form of these coefficients is given in the appendix. Thus, there are as many equations of the type (10) as inner points on the mesh. So far, the only approximation performed was to replace each derivatives with a central–difference formula. Otherwise, the system of equations (10) is consistent and has the same thermodynamical information as (4). For an useful application, the system of coupled algebraic equations (10) is huge; eg. there are 15,000 unknown in a mesh of 300 isotherms and 50 isochores. Due to the difficulties involving in (10), we relied on an iterative method of solution named as the SOR method. In this method, we gave an initial guess to all inner point $Z_{j,i}^{(0)}$ and we solved (10) using both the boundary conditions given for Z along two isotherms (j fixed at j_0 and J), and two isochores (i fixed at i_0 and I) and the speed of sound data. Calling $Z_{j,i}^{'(n+1)}$ to the solution of (10) for the $(n + 1) - -th$ iteration as a function of the four $Z^{(n+1)}$ neighbors already updated and the other four $Z^{(n)}$ neighbors without update, the new value for $Z_{j,i}^{(n+1)}$ is given by

$$Z_{j,i}^{(n+1)} = \omega Z_{j,i}^{'(n+1)} + (1 - \omega) Z_{j,i}^{(n)}, \tag{11}$$

where $1 < \omega < 2$ is a relaxation parameter used to accelerate the convergence rate. In our case, we optimized the value of ω after few trials. The convergence criterium used was that the rms change of the whole surface between two successive iteration were less than 1×10^{-5}.

4. RESULTS AND DISCUSSION

In this section we shown the results obtained for argon from the BC method, and we compared these results with those derived from the IV approach. In both cases, Z and $C_{V,m}$ were determined in the gaseous phase and the data required as an input data were taking from an empirical equation of state (eos), [9] which was taken as a reference surface. For both methods we covered the same region with $\tau \in [1, 4]$ and $\delta \in [0, 0.5]$; no errors on initial or boundary conditions neither on u were considered in these calculations. In the table, we summarized the information about the parameters required by the integration of each method. The initial guess for each inner point in the mesh was always the ideal gas solution; i.e. $Z_{j,i}^{(0)} = 0$ for all $j_0 < j < J$ and $i_0 < i < I$. This guess was proposed because in the gaseous phase, Z is close to the ideal–gas limit, and even near to the critical region, this guess seems to be an adequate choice. With these conditions, we compared the results of each method ($Z^{(\text{int})}$) against the values predicted by the eos ($Z^{(\text{eos})}$) and we plotted the deviations $\Delta Z = Z^{(\text{int})} - Z^{(\text{eos})}$ and $\Delta C_{V,m} = C_{V,m}^{(\text{int})} - C_{V,m}^{(\text{eos})}$ as 3–d surfaces in each case. The perspective and the scale of all these plots are the same and, all data plotted were smoothed by a cubic–spline interpolation. Although, the number of points generated for each method is different (for the IV were about 45,000 while for the BC were only about 15,000), we selected in all figures a similar amount of points evenly distributed, (about 2500 points) so the differences found between the two methods were not due to the different number of points integrated. The absolute deviations ΔZ are shown in figures 1 and 2, for the IV and the BC approach, respectively. We want to stress that in figure 2, ΔZ is zero along the border while in figure 1, ΔZ is only zero in two limits ($\tau = 1$ and $\delta = 0$). There are two main features that figure 1 shows, firstly there is main maximum at $\tau \sim 1.7$ which increases linearly with density and secondly, a ripples–formation developed at $\tau \sim 2.5$ and $\delta \sim 0.25$. The maximum in figure 1 can be explained in terms of an error in $(\partial Z/\partial T)$ at $\tau_0 + \Delta\tau$ rather than at the initial temperature τ_0 [7] but there is no thermodynamic explanation to the ripples though they are not greater than few parts per million of Z. In contrast, we found no systematic deviations in figure 2 and ΔZ is no larger than $\pm 3 \times 10^{-6} \cdot Z$.

Table 1. Parameters used in the numerical integration for the BC and IV method.

	$\Delta\tau$	$\Delta\delta$	ω	j_0	i_0	J	I
BC	1×10^{-2}	1×10^{-2}	1.75	100	0	399	49
IV	$(1/3) \times 10^{-3}$	1×10^{-2}					

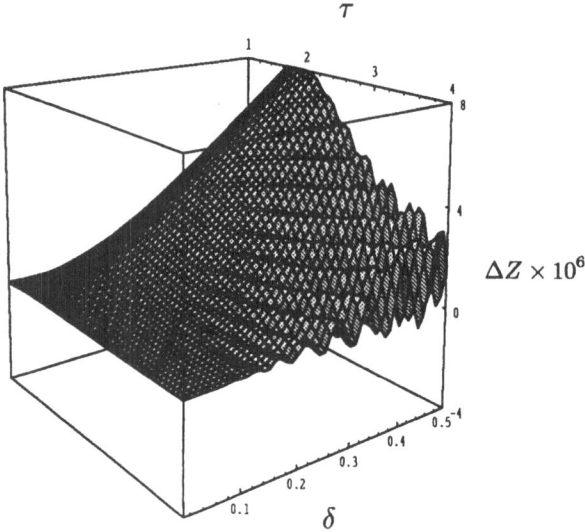

Figure 1.   Deviations ΔZ against eos in reference [9] from the IC method

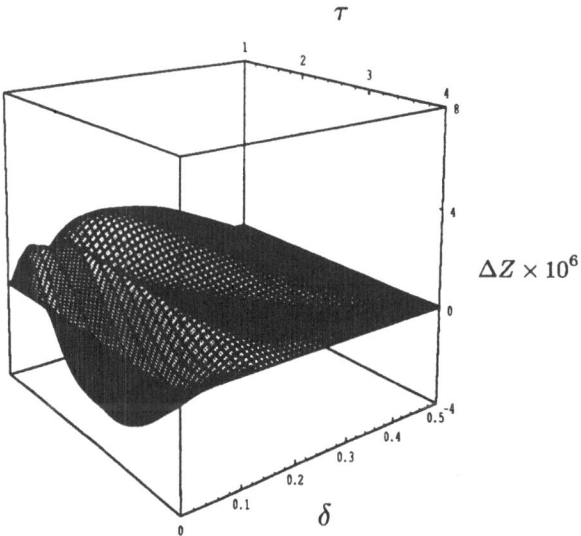

Figure 2.   Deviations ΔZ against eos in reference [9] from the BC method

In figures 3 and 4, we presented the results of $\Delta C_{V,\mathrm{m}}$. In the IV method, there is a well–defined region where ripples-formation is observed in both directions, along isotherms and isochores.

These ripples have been reported previously [8] and they were the actual reason to explore the BC approach. Here, the ripples only amounted

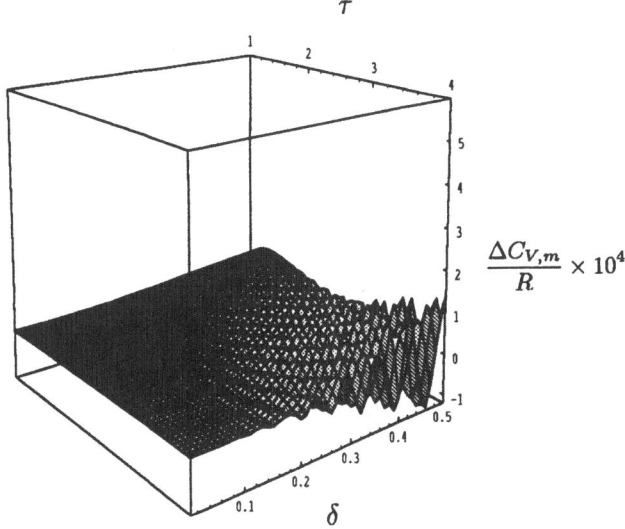

Figure 3. Deviations $\Delta C_{V,\mathrm{m}}$ against eos in reference [9] from the IC method

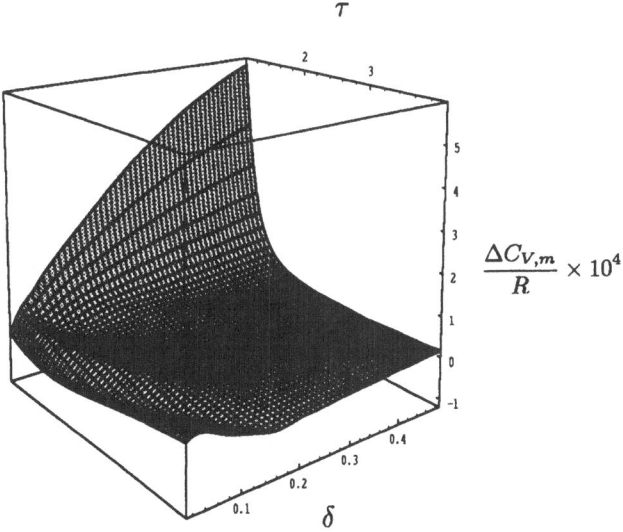

Figure 4. Deviations $\Delta C_{V,\mathrm{m}}$ against eos in reference [9] from the BC method

to $\pm 1 \times 10^{-4} \cdot R$ but in an integration with experimental data of u the situation may be worse; we know that interpolation of u may increase the ripples about hundred times larger. For the BC method this feature disappeared; no evidence of ripples is observed in figure 4. However, there is a maximum for $\Delta C_{V,\mathrm{m}}$ at the initial isotherm for which we have not found explanation. We suggest that this maximum would be reduced using a finer mesh but then the amount of inner points would growth rapidly and the calculations needed may compromise the results for a larger region in the (τ, δ) plane.

5. CONCLUSIONS

We have shown two quite accurate methods for deriving Z and $C_{V,\mathrm{m}}$ from u. By and large, both methods, IV and BC, gave the same value of the rms of ΔZ, but in one of them (BC) no systematic errors were found. In the case of $C_{V,\mathrm{m}}$, the BC method seems to be a real improvement, even when the results for $C_{V,\mathrm{m}}$ derived from the IC method have been much better than any direct measurement of this property. Thus, accurate and precise values of $C_{V,\mathrm{m}}$ can be derived from speed of sound measurements supplemented with minimum information of pVT data.

Acknowledgments

This work has been supported by grant 400200–5–35335-E of Conacyt. One of us (D.J.) wish to thank Conacyt for financial support.

Appendix

The coefficients b_l of the cubic equation of $Z_{j,i}$, equation (10), are the following

$$
\begin{aligned}
b_2 =& -\{j[(j+1)Z_{j+1,i} + (j-1)Z_{j-1,i} + 4jF_{j,i} - 2ji(Z_{j,i+1} - Z_{j,i-1})] + \\
& i[(i+1)Z_{j,i+1} + (i-1)Z_{j,i-1} - (F_{j,i+1} - F_{j,i-1})/2 - 2ji(Z_{j+1,i} - Z_{j-1,i})] - \\
& i[(j/2)(Z_{j+1,i+1} - Z_{j-1,i+1} - Z_{j+1,i-1} + Z_{j-1,i-1})]\}/[2(j^2 + i^2)], \\
b_1 =& \{2j[F_{j,i} - (i/2)(Z_{j,i+1} - Z_{j,i-1})] \times \\
& [(j+1)Z_{j+1,i} + (j-1)Z_{j-1,i} + jF_{j,i} - (ij/2)(Z_{j,i+1} - Z_{j,i-1})] + \\
& [(ij)(Z_{j+1,i} - Z_{j-1,i})] \times \\
& [(ij/2)(Z_{j+1,i} - Z_{j-1,i}) - (i+1)Z_{j,i+1} - (i-1)Z_{j,i-1} + (F_{j,i+1} - F_{j,i-1})/2] - \\
& i[(j/2)(Z_{j+1,i+1} - Z_{j-1,i+1} - Z_{j+1,i-1} + Z_{j-1,i-1}) + Z_{j,i+1} - Z_{j,i-1}] \times \\
& [F_{j,i} - (i/2)(Z_{j,i+1} - Z_{j,i-1}) - (j/2)(Z_{j+1,i} - Z_{j-1,i})]\}/[2(j^2 + i^2)], \\
b_0 =& -\{j[F_{j,i} - (i/2)(Z_{j,i+1} - Z_{j,i-1})]^2[(j+1)Z_{j+1,i} + (j-1)Z_{j-1,i}] + \\
& i[(j/2)(Z_{j+1,i} - Z_{j-1,i})]^2[(i+1)Z_{j,i+1} + (i-1)Z_{j,i-1} - (F_{j,i+1} - F_{j,i-1})/2] + \\
& i[(j/2)(Z_{j+1,i+1} - Z_{j-1,i+1} - Z_{j+1,i-1} + Z_{j-1,i-1}) + Z_{j,i+1} - Z_{j,i-1}] \times \\
& (j/2)[Z_{j+1,i} - Z_{j-1,i}][F_{j,i} - (i/2)(Z_{j,i+1} - Z_{j,i-1})]\}/[2(j^2 + i^2)].
\end{aligned}
$$

References

[1] M.R. Moldover, J.B. James and M. Greenspan, *J. Acoust. Soc. Am.* **79** (1985) 253.

[2] J.L. Daridon, B. Lagourette and P. Xans, *Fluid Phase Equilibria* **100** (1994) 269.

[3] T.C. Dayton, S.W. Beyerlein and A.R.H. Goodwin, *J. Chem. Thermodynamics* **484** (1999) 847.

[4] S. Detour, J.L. Daridon, and B. Lagourette, *Inter. J. Thermophysics* **100** (2000) 173.

[5] J.J. Hurly, *Inter. J. Thermophysics* **21** (2000) 805.

[6] M.R. Riazi and G.A. Mansoori, *Fluid Phase Equilibria* **90** (1993) 251.

[7] A.F. Estrada-Alexanders, J.P.M. Trusler, and M.P. Zarari, *Inter. J. Thermophysics* **16** (1995) 663.

[8] A.F. Estrada-Alexanders and J.P.M. Trusler, *Inter. J. Thermophysics* **17** (1996) 1325.

[9] R.B. Stweart, R.T. Jacobsen and J.H. Becker, *Center for Applied Thermo. Studies Report No. 81-3*, Univ. of Idaho, 1981.

EFFECTIVE POTENTIALS AND SECOND VIRIAL COEFFICIENT FOR POLAR FLUIDS

F. del Rio and E. Ávalos

Departamento de Física, Universidad Autónoma Metropolitana
Av. San Rafael Atlixco 186, Col. Vicentina, Iztapalapa, México, D.F., CP 09340 México

Abstract   Using the Approximate Non-Conformal theory (ANC), effective potentials for model substances with dipolar interactions are built and shown to reproduce the data of their second virial coefficient. The systems considered are the Stockmayer potential and a modified Stockmayer in which the spherical Lenanrd-Jones part is replaced by a more realistic function of the Kihara type.

Keywords: Effective potentials, second virial coefficient, polar fluids, Stockmayer potential

1.   INTRODUCTION

The ANC theory was proposed as a rigorous extension of the principle of corresponding states for gases. Here we present a brief description of its main features. The interested reader may consult the original references in [2]. The ANC theory introduces the softness of the potential, s, as a new molecular parameter to account for the form of the potential profile. Besides the shape parameter s, this theory uses an effective spherical potential with minimum at r_m of depth ϵ, which reproduces the second virial coefficient $B(T)$ of a wide class of non-conformal substances and their binary mixtures. The ANC potential is

$$\varphi_{ANC}(z, s; \epsilon) = \epsilon \left\{ \left(\frac{1-a}{\zeta(z,s) - a} \right)^{12} - 2 \left(\frac{1-a}{\zeta(z,s) - a} \right)^{6} \right\}, \quad (1)$$

where $\zeta(z,s) = (1 + (z^3 - 1)/s)^{1/3}$, $z = r/r_m$, and a is a constant. The potential $\varphi_1(z) = \varphi_{ANC}(z, s = 1)$ is the ANC reference potential and we choose $a = 0.0957389$ so that $\varphi_1(z)$ is closely conformal to the true

Developments in Mathematical and Experimental Physics,
Volume B: Statistical Physics and Beyond
Edited by Macias *et al.*, Kluwer Academic/Plenum Publishers, 2003

pair potential for argon. Varying the softness s in $\varphi_{\mathrm{ANC}}(z, s)$ changes the shape of the potential profile: decreasing s makes the potential well narrower, i.e. steeper or harder, and viceversa, increasing s makes the well wider. When the substance in consideration has a spherical symmetric potential $\varphi(z)$ with equal softness on its attractive and repulsive parts, the ANC theory provides us with an exact linear relation between the reduced second virial coefficients of the substance $B^*(T^*)$ and of the reference $B_1^*(T^*)$ [2]:

$$B^*(T^*) = 1 - s + sB_1^*(T^*), \qquad (2)$$

where $B^*(T^*) = B(T)/(2\pi r_{\mathrm{m}}^3/3)$ and $T^* = kT/\epsilon$, with constant r_{m} and ϵ. When the intermolecular pair potential is angle dependent, (2) is obtained as an approximation. The ANC theory has been applied successfully to a wide variety of real and model substances in the gas phase and, more recently, to dense fluids [1]. For nonpolar fluids, the ANC theory with constant parameters $(\epsilon, r_{\mathrm{m}}, s)$ has been able to give very accurate effective potentials for chain molecules as elongated as C_8H_{18} and C_7F_{16} [5]. The constant–parameter ANC theory has provided a reasonable approximation only for slightly-polar quasi–spherical substances such as NO and CO but has failed for strongly polar substances such as alcohols.

Here we report preliminary results directed to a more accurate treatment of polar substances within the ANC theory. It is shown that when the interaction potential consists of a spherical part plus a permanent dipole–dipole term, the corresponding effective ANC potential will have parameters $(\epsilon, r_{\mathrm{m}}, s)$ that depend on both the temperature T and the dipolar moment μ. It is also shown that this T–dependent ANC potential reproduces accurately $B(T)$ for the cases considered.

2.   MODEL POLAR FLUIDS

We calculate here the effective ANC potential for two polar model systems, namely, the well–known Stockmayer (SM) potential and a modified Stockmayer potential (MSM), which is made by adding a dipolar term to the ANC reference $\varphi_1(z)$.

2.1.   STOCKMAYER POTENTIAL

In reduced units, the Stockmayer (SM) potential is

$$\varphi_{\mathrm{SM}}(x, \Omega, \widetilde{\mu}) = \varphi_{\mathrm{LJ}}(x) + \varphi_{\mathrm{dd}}(x, \Omega, \widetilde{\mu}) \qquad (3)$$

with

$$\varphi_{LJ}(x) = 4\epsilon_{LJ}\left\{\frac{1}{x^{12}} - \frac{1}{x^6}\right\} \text{ and } \varphi_{dd}(x,\Omega,\widetilde{\mu}) = -\epsilon_{LJ}\frac{\widetilde{\mu}^2}{x^3}g(\Omega), \qquad (4)$$

where

$$g(\Omega) = g(\theta_1,\theta_2,\phi_2 - \phi_1) = 2\cos\theta_1\cos\theta_2 - \sin\theta_1\sin\theta_2\cos(\phi_2 - \phi_1),$$

$x = r/\sigma_{LJ}$, r is the center–to–center distance, $\widetilde{\mu} = \mu/\sqrt{\epsilon_{LJ}\sigma_{LJ}^3}$, whereas θ_i and ϕ_i are the zenithal and azimuthal angles of the dipole on molecule i. The subscript in ϵ_{LJ} and σ_{LJ} means that these quantities are the same as in the single Lennard–Jones potential.

We start by following the procedure due to Hirschfelder et al. [3] The SM second virial coefficient is known from previous authors who reduced it with σ_{LJ}^3 instead of r_m^3, hence we let $\widetilde{B}(T) = B(T)/\frac{2}{3}\pi\sigma_{LJ}^3$, so that $\widetilde{B}(T) = x_{meff}^3 B^*(T^*)$ and is given by

$$\widetilde{B}(T) = \frac{1}{8\pi}\int_0^\infty x^2 dx \int d\Omega \frac{\partial}{\partial x}e^{-\varphi_{SM}(x,\Omega)/kT} \qquad (5)$$

Factorizing $\exp(-\varphi_{SM}/kT) = \exp(-\varphi_{LJ}(x)/kT) \times \exp(\varphi_{dd}(x,\Omega)/kT)$, expanding the dipolar factor $\exp(\varphi_{dd}/kT)$ and integrating over the angles one finds

$$\int d\Omega \exp(\varphi_{dd}/kT) = \sum_{m=0}^\infty \left\{\frac{1}{(2m)!}\left[\widetilde{\mu}^2/\widetilde{T}\ x^3\right]^{2m}G_m\right\} \qquad (6)$$

where $\widetilde{T} = kT/\epsilon_{LJ}$ and $G_m = (1/8\pi)\int d\Omega\ g^{2m}(\Omega)$ may be calculated analytically for m finite. At this point we depart from Hirshfelder et al. procedure and define the auxiliary spherical potential $\widetilde{\varphi}_{dd}$ by

$$\widetilde{\varphi}_{dd}(x,\widetilde{T},\widetilde{\mu}) = -\epsilon_{LJ}\widetilde{T}\ln\sum_{m=0}^\infty \left\{\frac{1}{(2m)!}\left[\widetilde{\mu}^4/\widetilde{T}^2 x^6\right]^m G_m\right\}, \qquad (7)$$

Adding this to $\varphi_{LJ}(x)$ shows that the Stockmayer potential may be replaced by a spherical potential which depends on the parameters $\widetilde{T}$ and $\widetilde{\mu}$:

$$\varphi_{SM}(x,\Omega,\widetilde{\mu}) \to \varphi_{SM}^{eff}(x,\widetilde{T},\widetilde{\mu}) = \varphi_{LJ}(x) + \widetilde{\varphi}_{dd}(x,\widetilde{T},\widetilde{\mu}) \qquad (8)$$

we notice that expanding the logarithm in (7) the term $m = 1$ gives rise to the Keesom potential. [6]

We now proceed with the ANC procedure and look for the ANC potential $\varphi_{ANC}(z,s)$ that corresponds to a given $\varphi_{SM}^{eff}(x,\widetilde{T},\widetilde{\mu})$. To do that we must find the ANC effective parameters r_{meff}, ϵ_{eff} and s; the first two expressed in LJ units are $x_{\text{meff}} = r_{\text{meff}}/\sigma_{\text{LJ}}$ and $\widetilde{\epsilon}_{\text{eff}} = \epsilon_{\text{eff}}/\epsilon_{\text{LJ}}$. First, we notice that due to the influence of the dipolar interaction, $\varphi_{\text{SM}}^{\text{eff}}$ has a minimum deeper than $\varphi_{\text{LJ}}(x)$ and its position is shifted. Hence for given values of $\widetilde{\mu}$ and $\widetilde{T}$ we re–scale $\varphi_{\text{SM}}^{\text{eff}}$ by the values of the position of its minimum x_{meff} and its depth ϵ_{eff} so that $\varphi_{\text{eff}}(x/x_{\text{meff}})/\epsilon_{\text{eff}}$ has a value -1 at $x/x_{\text{meff}} = 1$. This re–scaling gives x_{meff} and ϵ_{eff}; further, comparing the profiles of the re–scaled potentials $\varphi_{\text{SM}}^{\text{eff}}(x/x_{\text{meff}})/\epsilon_{\text{eff}}$ for $0.3 \leq \widetilde{T} \leq 10$ and $\widetilde{\mu} \leq 1.0$ we found that they are conformal, i.e. all af them have the same value of s as the LJ potential (which corresponds to $\widetilde{\mu} = 0$ or $\widetilde{T} \to \infty$). Therefore, the appropriate ANC potential (1) will have softness $s = s_{\text{LJ}} \simeq 1.12$, depth $\epsilon_{\text{eff}}(\widetilde{T}, \widetilde{\mu})$ and minimum at $x_{\text{meff}}(\widetilde{T}, \widetilde{\mu})$, i.e.

$$\varphi_{\text{ANC}}(z, s_{\text{LJ}}; \widetilde{\epsilon}_{\text{eff}}) \cong \varphi_{\text{SM}}^{\text{eff}}(x, \widetilde{T}, \widetilde{\mu}) \qquad (9)$$

where now $z = r/r_{\text{meff}} = x/x_{\text{meff}}(\widetilde{T})$.

Hence, from (2), the ANC theory predicts that $\widetilde{B}\left(\widetilde{T}, \widetilde{\mu}\right)$ is given by

$$\widetilde{B}(\widetilde{T}) = x_{\text{meff}}^3(\widetilde{T}, \widetilde{\mu}) \left\{ 1 - s_{\text{LJ}} + s_{\text{LJ}} \, B_1^*\left(T^*\right) \right\}, \qquad (10)$$

where $T^* = kT/\widetilde{\epsilon}_{\text{eff}}$.

Here we present results for $\widetilde{\mu} = 1.0$. The minimum of $\varphi_{\text{eff}}(x, \widetilde{T}, 1)$ was calculated for $0.3 \leq \widetilde{T} \leq 10$ in steps of 0.1 and the results were fitted by functions of the form $\widetilde{\epsilon}_{\text{eff}}(\widetilde{T}) = C_1 + C_2/\widetilde{T} + C_3/\widetilde{T}^2$ and $x_{\text{meff}}(\widetilde{T}) = C_1' + C_2'/\widetilde{T} + C_3'/\widetilde{T}^2$ to obtain the coefficients C_i and C_i'. By using these fits we can predict $\widetilde{B}(\widetilde{T}, \widetilde{\mu} = 1.0)$ using (10). Figure 1 shows the prediction of ANC theory for $\widetilde{B}(\widetilde{T})$ compared with the results obtained from direct integration following Hirschfelder et al. [3] The good agreement between them shows the high accuracy of the effective potential as given by the ANC theory.

2.2.   MODIFIED STOCKMAYER POTENTIAL

The second potential considered, $\varphi_{\text{MSM}}(z, \Omega, \mu^*)$, is built by adding a reference ANC potential, $\varphi_{\text{ANC}}(z, s = 1)$ with depth ϵ_1 and minimum at r_{m1}, i.e. $z = r/r_{\text{m1}}$, and a dipolar term that is written now as $\varphi_{\text{dd}}(z, \Omega, \mu^*) = \epsilon_1 \mu^{*2} g(\Omega)/z^3$, where $\mu^* = \mu/\sqrt{\epsilon_1 r_{\text{m1}}^3}$. There are no calculations reported for $B(T)$ of this potential, hence we also present the results of the direct procedure. Integrating to calculate the second virial potential, in a way similar to that already described for the Stockmayer

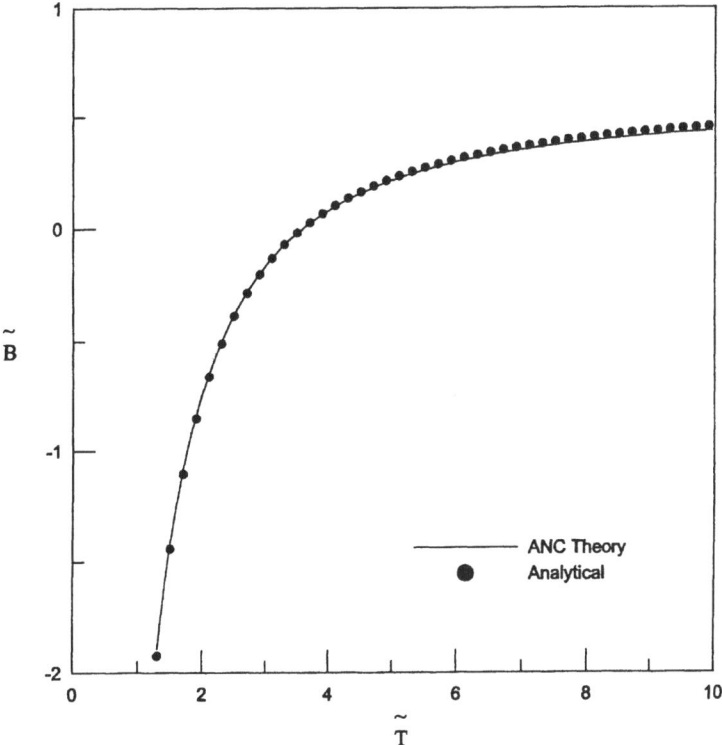

Figure 1. Result of ANC second virial coefficients as a function of temperature for $\widetilde{\mu} = 1.0$. Points are the results from direct analytical integration, eq. (11) and line the result of ANC theory.

case, we get

$$
B^*\left(T^*, \mu^*\right) = -\sum_{n=0}^{\infty}\sum_{k=0}^{k\leq\frac{n}{2}}\sum_{l=0}^{\infty}\frac{2^{n-2k}}{n!}\frac{a^l}{(1-a)^{6k+l}}\left(\frac{1}{T^*}\right)^{\frac{6n+6k-l}{12}} \tag{11}
$$
$$
\times\binom{n}{2k}\left\{\begin{matrix}6k\\l\end{matrix}\right\}\mu^{*4k}G_k\{(1-a)^3\frac{1}{4}\left(\frac{1}{T^*}\right)^{\frac{1}{4}}
$$
$$
\times\Gamma6n-6k+l-312+(1-a)^2a\frac{1}{2}\left(\frac{1}{T^*}\right)^{\frac{1}{6}}
$$
$$
\times\Gamma6n-6k+l-212+(1-a)a^2\frac{1}{4}\left(\frac{1}{T^*}\right)^{\frac{1}{12}}
$$
$$
\times\Gamma6n-6k+l-112\},
$$

where $\begin{pmatrix} n \\ 2k \end{pmatrix}$ is the coefficient of the binomial Newton expansion, $\Gamma(\cdot)$ is the gamma function, G_k is the same as above and

$$\begin{Bmatrix} 6k \\ l \end{Bmatrix} = (-1)^l \frac{(6k+l-1)!}{l!(6k-1)!}. \tag{12}$$

Analogously to the Stockmayer potential, integrating the expansion of $\exp(\varphi_{dd}(z,\Omega,\mu^*)/kT)$ over the angles we find again that

$$\int d\Omega \exp(\varphi_{dd}/kT) = \exp\left[\widetilde{\varphi}_{dd}(z,T^*,\mu^*)/kT\right] \tag{13}$$

so that $\varphi_{MSM}(z,\Omega,\mu^*)$ is replaced by an effective one whose parameters ϵ_{eff} and z_{eff} again depend on T^* and μ^*:

$$\varphi_{MSM}(z,\Omega,\mu^*) \rightarrow \varphi_{MSM}^{eff}(z,T^*,\mu^*) = \varphi_1(z) + \widetilde{\varphi}_{dd}(z,T^*,\mu^*) \tag{14}$$

As done above, we re-scaled $\varphi_{MSM}^{eff}(z,T^*,\mu^*)$, at $0.3 \leq T^* \leq 10$ and $\mu^* \leq 1.0$, to find the position of its minimum z_{meff} and its depth $\widetilde{\epsilon}_{eff}$. As in the Stockmayer case, the profiles of the re-scaled potential $\varphi_{MSM}^{eff}(z/z_{eff})/\epsilon_{eff}$ match each other to a very good approximation, so that the presence of the dipolar interaction does not affect significantly the softness, $s = 1$, of the potential. Again we can fit these parameters $\epsilon_{eff}^*(T^*,\mu^*) \equiv \epsilon_{eff}(T^*,\mu^*)/\epsilon_1$, and $z_{meff}^*(T^*,\mu^*) \equiv r_{meff}(T^*,\mu^*)/r_{m1}$ with appropriate functions of T^* and μ^*. Thus $\varphi_{MSM}^{eff}(z/z_{eff},T^*,\mu^*)/\epsilon_{eff}$, will have $s = 1$ and parameters z_{eff} and ϵ_{eff} that depend on T^* and μ^*. Hence, in this case, the appropriate ANC potential will be

$$\varphi_{ANC}(z,s=1;\epsilon_{eff}^*) \cong \varphi_{MSM}^{eff}(z/z_{eff},T^*,\mu^*) \tag{15}$$

Hence the ANC theory predicts that $B^*(T^*;\mu^*)$ is given by

$$B^*(T^*,\mu^*) = z_{meff}^3(T^*,\mu^*)\{1 - s + s\, B_0^*(T^*/\epsilon_{eff}^*(T^*,\mu^*))\} \tag{16}$$

The points in figure 2 show the numerical results from direct analytical integration, eq. (11), with $n_{max} = 50$ and $l_{max} = 30$, for $\mu^* = 0.5$, 0.75 and 1.0, and for $T^* \in [0.3, 10)$, and the continuous line is the prediction of the ANC theory. As in the previous case, the ANC potential is seen to be a highly accurate effective potential for the MSM case.

3. CONCLUSIONS

The ANC theory has been shown to give very accurate effective potentials for a couple of dipolar systems, namely, the Stockmayer potential

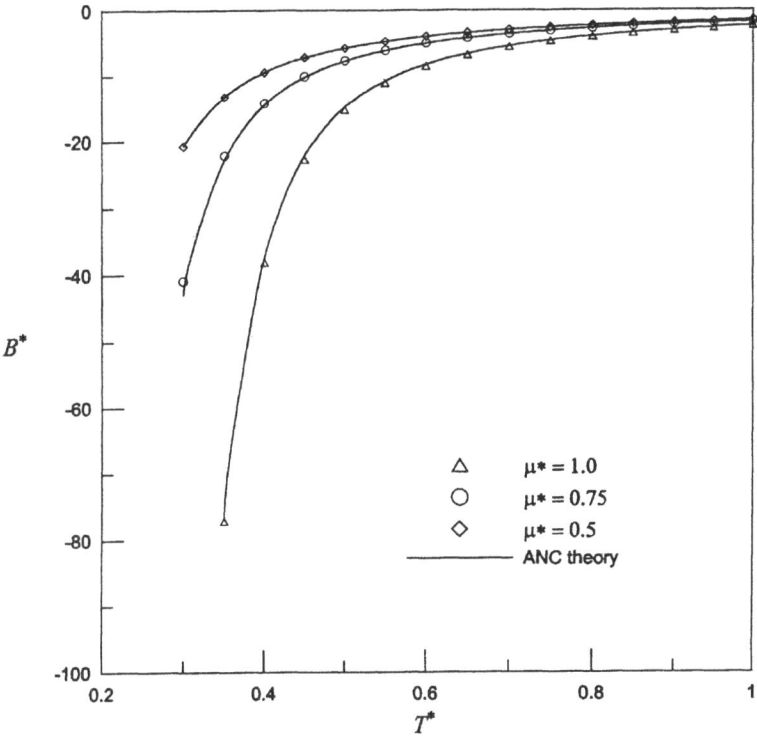

Figure 2. ANC reduced second virial coefficients as a function of reduced temperature (lines) in comparison with the results obtained from direct integration following Hirschfelder et al. [3] (symbols)

and the modified Stockmayer (which contains a $s = 1$ ANC potential instead of the LJ one). In the first case, the reported method reproduces $B(T)$ in very good agreement with the values reported by Maitland *et al* [4]. In the MSM case we succeeded in reproducing the results calculated by direct analytical integration. The main point is that, in order to apply the ANC theory to these polar gases, the ANC parameters ϵ and r_{m} have to depend on T and μ. The method is going to be applied to other MSM systems whose spherical part has $s < 1$ which corresponds to most substances thus far studied.

Acknowledgments

E.A. thanks Conacyt (México) for the scholarship granted for Ph.D. studies.

References

[1] O. Guzmán, F. del Río, *The Journal of Physical Chemistry* **B105** (2001) 8220.

[2] F. del Río, J. E. Ramos, I. A. McLure, *The Journal of Physical Chemistry* **B102** (1998) 10568.

[3] J. O. Hirschfelder, C. F. Curtiss and R. B. Bird, *Molecular Theory of Gases and Liquids*, 1954, (John Wiley & Sons, New York, 1954).

[4] G.C. Maitland, M. Rigby, E.B. Smith, W. A. Wakeham, *Intermolecular Forces* (Clarendon Press, 1981).

[5] J. E. Ramos, F. del Río, I. A. McLure, *J. Chem. Phys.* **2** (2000) 2731.

[6] R.H. French, *Journal of the American Ceramic Society* **83** (2000) 86.

LEOPOLDO GARCIA–COLIN SCHERER: BRIEF BIOGRAPHY

Eduardo Piña

Leo, as he is known among some of his foreign colleagues, was born in Mexico in 1930, but with strong roots in Spain and Germany, and sharing the Spanish and English as mother languages.

He studied Chemistry and Physics simultaneously at two different faculties of the University of Mexico finishing his Bachelor in Science studies in 1954. His thesis dissertation to obtain his B.Sc. degree in Chemistry dealt with the thermodynamic properties of D_2 and HD as part of a project whose leader was one of Mexicos's foremost scientist, Alejandro Medina. The project was the construction of a nuclear reactor moderated with heavy water, which finally, was never constructed. He afterwards completed his Ph.D. work in Theoretical Physics at the University of Maryland in 1960 under Elliott Montroll's guidance. There he began his dedicated career as a teacher and researcher and started an important research program on the broad area of Statistical Mechanics, Kinetic Theory, Irreversible Thermodynamics, Critical Phenomena, Chemical Physics, and other related subjects.

By personal interest, my favorites in the hundreds of papers he has published are those on Critical Phenomena published with former students of him, others on the frontier problems of Kinetic Theory which were probably unfinished and included many years of collaboration with Melville Green and many other scientists. Also I prefer those on the Modern Foundations of Irreversible Thermodynamics. But my taste is based by personal preferences, and I do not pretend to make a fair selection.

Returning to Mexico in 1960 he faced the very hard duty of constructing from nothing, totally new research groups in the field of Sciences he cultivated, creates and loves, and he spent many of his time, energy, and insight in the development of high level Educative Institutions, and Government Laboratories with a firm scientific basis. The first task was to optimize the scientific formation of scientists, mixing the selection and preparation of students, with a program of graduate studies outside Mexico in selected places abroad, including the best universities of the United States, England, Holland, Germany, Belgium, etc. The role of Leo was essential through his personal scientific friends and friends of friends, since through their disposition of accepting students contributed to the enhancement of Statistical Physics in Mexico.

This fact is now fully acknowledged in Mexico. Prof. Garcia–Colin has been recognized as the founder of Statistical Physics in Mexico and has been honored with the highest level of National Researcher namely, he was accepted as a member of the select COLEGIO NACIONAL, hosting us today. Moreover, he was honored with the *National Prize of Sciences and Arts*, and also with the prize of the Mexican National Academy of Sciences; making him the first *Distinguished Professor* at the *Universidad Autónoma Metropolitana* of Mexico, where he has taught and done research the last twenty eight years; and appointing him *Doctor Honoris Causa* in two important universities in the country.

Along many years, beside the fields s previously mentioned, his interest in Science has had important incursions on General Relativity, Air Pollution Problems, Politics of Science, and Educational Research.

Some scientific textbooks have been authored by him, published by serious scientific editorials of Mexico for the service of the Spanish–speaking students.

This gigantic labor that can be only partially understood and appreciated in other countries, has been partaken with an intense research activity resulting in many papers published in the several International Journals of Physics and Physical Chemistry and is well known to all of those interested in current science. He has maintained a continuous correspondence with many scientists and does frequent scientific travels all around the world. He is an active member of various prestigious International Societies.

A sporting man, he is a regular amateur member of the jai alai, the basket Spanish ball game, which he inherited from his Father, an important professional of this game. But he has been also a practitioner of many other sports like racket fronton, marathon running and swimming.

He is a family man, with a beautiful family. He loves the music of the best composers and he will convince his interlocutors of this pleasure. He reads on so many matters, so many books, and so many journals.

I apologize for not being more clear on the importance of all his achievements, or the intimate collaboration with all those around him. Today I have not enough time to extend myself more on his achievements. In the future many other Mexican historians will recognize and give a fair and full explanation of my speech.

NICHOLAS G. VAN KAMPEN: BRIEF BIOGRAPHY

Rosalio F. Rodriguez

I am greatly honored and very pleased to present PROF. NICHOLAS G. VAN KAMPEN, who is the first awardee of the LEOPOLDO GARCÍA-COLÍN SCHERER MEDAL. A prize that bears the name of an outstanding Mexican scientist who started, developed and continues consolidating and expanding Statistical Physics in this country.

For more than half a century Nico van Kampen has been a pioneering force in theoretical physics and statistical mechanics. He has been an outstanding theoretical physicist ever since the publication of the results of his Ph. D. thesis in physics entitled CONTRIBUTIONS TO THE QUANTUM THEORY OF LIGHT SCATTERING, in the Proceedings of the Danish Academy of Sciences. He developed his dissertation under the advice of HANS KRAMERS in the university of Leyden and in the Niels Bohr Institute in Copenhagen, and obtained his Ph. D. degree in physics in 1951. In his thesis he showed how to overcome the singularities which arise in many quantum scattering processes. These results were essential to give the final development to Kramers's ideas which later on led to the methods of *renormalization*.

His "education" in statistical physics was accomplished in the best tradition of the Dutch School of Statistical Mechanics since his professor Kramers, was a former student of Ehrenfest whom, in turn, had been student of Boltzmann in Vienna.

In his early period of work in 1953, Nico became interested in the statistical mechanics of irreversible processes when he joined the group of PROFESSOR DE GROOT, the successor of Kramers in Leyden, and who introduced him into the field of Irreversible Processes.

In this early period in this field he addressed topics such as the quantum theory of the statistical mechanics of nonequilibrium processes, whose formulation was unsatisfactory because the familiar picture of eigenvalues and eigenstates was inappropriate for macroscopic systems.

In 1955 Nico moved to the University of Utrecht, where much experimental work on fluctuations had been done and where he developed one of his major lines of research, namely, that of the theoretical analysis of nonlinear fluctuations and of the stochastic treatment of noise in physical systems. In this context he clarified many fundamental issues concerning the use and proper place of stochastic processes in physics. He also showed how the theory of random events is spread out over the entire scientific literature from mathematics to biology. In this context, one of his most important contributions was the introduction of a systematic expansion method of the master equation. This method has clarified many misconceptions and fundamental issues in stochastic processes and has greatly expanded the applicability of Markovian processes to numerous systems in physics and chemistry.

His research on stochastic processes between the years 1955 and 1980 produced a long string of epoch making papers discussing a variety of controversial issues in the field. His powerful lines of thought, deep criticism and rigorous logical thinking addressed topics such as the *"use and abuse of the Langevin approach"*, *"the Ito–Stratonovich dilemma"*, *"the fluctuations in the Boltzmann equation"* or *"the stochastic behavior of quantum systems"*.

As is evident from the quality of his review articles on stochastic processes and other fields, he has a wonderful knowledge of the literature on this subject, and many of his views and contributions on these topics where collected in his very well known book: *"Stochastic Processes in Physics and Chemistry"*, first published in 1981 and revised and enlarged in 1992. This book is a classic in the field and has guided many graduate students and scientists through the complexities of research in different topics of fluctuations and nonequlibrium statistical mechanics.

Another field where he has made important contributions, was that of plasma physics, where in collaboration with Ubo Felderhof, he developed methods to derive the modes of the linear Vlasov equation and which are now called the *"van Kampen's modes"*.

He has been a visitor in several Universities like Aachen, in Germany, and Minneapolis, Howard's and Texas in the United States. It was precisely in the University of Texas at Austin in 1974, where he became interested on some aspects of the *foundations of quantum mechanics*, in particular, in the so-called *measurement theory of quantum mechanics*. To avoid the standard treatment of the problem which seemed to him so remote from reality, he constructed a simple model which contained all the essential elements of the problem. One of these elements was the essential relation of microscopic events to the macroscopic world which he had considered before within the context of large systems in quantum statistical mechanics. In this context, he is specially critical about those who try to endow quantum mechanics with some kind of *"mysticism"*. Argues van Kampen: *"quantum mechanics is a perfectly logical and coherent physical theory, which can be understood rationally. The mysticism is theirs."* His analysis and points of view on this issue were published in a series of papers and his main results appeared in a paper under the title *"Ten theorems about quantum mechanical measurements"* in 1988.

Among his widespread interests, and apart from his scientific writings, which stand by themselves, Nico van Kampen has also written extensively in a different category, namely, that of essays and miscellaneous writings. Recently, Paul Meijer has edited the book *"Views of a Physicist"*, which contains an extensive collection of Nico's essays on different subjects, most of which were not accessible in English. This collection of essays cover a variety of topics which range from writings for special occasions, such as invited lectures, speeches, popular–science writings, book reviews or obituaries. But it also includes several of his well known fundamental and incisive critical essays. For example, the one on the over–simplification of the statistical mechanical explanation of Ohm's law by Linear Response Theory; or the one on the usual, but erroneous, explanation of the Third Law of Thermodynamics on the basis of the non–degeneracy of the ground state.

This book also includes beautiful essays on great physicists, like Copernicus, Smoluchowski, Wigner and Kramers. Perhaps some of the most delightful essays are those about his *Recollections of Kramers*, his former teacher, and those that include his subtle, clear and first hand analysis of Kramer's work and achievements.

As Paul Meijer says in the preface of this book: *"these essays are sometimes philosophical, often critical and almost always enjoyable for their style alone. His style is incisive but not derogatory and often playful"*.

Indeed, as his scientific papers, these essays are also clear, accurate, critical and carefully written. Assets that are reflected over and over in Nico's prominent and brilliant scientific career.

As his nephew and 1999 Nobel–prize winner Gerard t'Hooft writes in the prologue of this book: *"Here comes van Kampen to show up the charlatans"*. Yes, indeed, here comes Nico, and we do hope that for many years to come! Thank you.

Index

alkylcyanobiphenyl, 228, 232
amphiphilic molecules, 125
ANC
    potential, 253
    theory, 253

biomembranes, 125, 126, 130
black-body radiation, 187, 188, 192, 194
Bose
    function, 189, 190
Bose-Einstein
    condensation, 187, 191
bosonic gas, 187, 188
brain, 99, 104
Brillouin
    components, 198
    spectrum, 211
Brownian
    dynamics, 88, 97
    motion, 15
    motors, 71, 82
    particles, 63, 65–68, 71, 82

catalysis, 15
charge density wave, 171, 173
Clausius-Mossotti
    relation, 162
colloid, 28
    dipolar, 28, 32
    dipolar spherical, 28
colloidal
    charged particle, 28
    dispersion, 35
    fluid, 15, 16
    mixture, 3, 4, 45, 46
    particle, 3, 4, 12, 13, 27, 28, 32, 45
        velocity, 32
    suspension, 27
    system
        dynamics, 45, 46, 55
        quasi-bidimensional, 12
    transport properties, 70
compression factor, 243, 244
critical solution, 135, 136

cytochrome, 87, 88, 90

diffusion, 87, 90
    coefficient, 31, 52, 96
        charged particles, 55
        long-time, 37
        short-time, 29, 32
    collective, 29, 39, 63, 68, 69
    constant, 16, 24, 96
    polyions, 42
    tensor, 29, 90
dynamic structure factor, 197–199, 205, 207
dynamical
    critical exponent, 216, 218–220, 223

effective
    chirality, 150
    electric
        permittivity, 147, 151, 153, 166, 169
    electrical
        conductivity, 150
    magnetic
        permeability, 147, 151, 152, 158, 166, 167
        susceptibility, 149, 166, 168, 169
    medium, 147, 149, 150, 152, 154, 158, 162, 169
        optical coefficients, 155
        theory, 155, 169
    pair potentials, 4, 5, 11
        charged macroparticles, 8, 11
        colloidal particles, 4, 12
        colloidal systems, 3
        DLVO, 9
        macroparticles, 12
    potential, 253, 254, 256, 258
    thermal
        conductivity, 150
    viscosity, 150
electric susceptibility, 159
electroencephalograms, 99
electrophoretic mobility, 27, 45, 46, 53